# Hubble Space Telescope

Nasa's Plans for a Servicing Mission

*(The Legacy of the World's Most Famous Telescope)*

## John Solomon

Published By **Oliver Leish**

## John Solomon

All Rights Reserved

*Hubble Space Telescope: Nasa's Plans for a Servicing Mission (The Legacy of the World's Most Famous Telescope)*

## ISBN   978-1-77485-777-9

No part of this guidebook shall be reproduced in any form without permission in writing from the publisher except in the case of brief quotations embodied in critical articles or reviews.

Legal & Disclaimer

The information contained in this ebook is not designed to replace or take the place of any form of medicine or professional medical advice. The information in this ebook has been provided for educational & entertainment purposes only.

The information contained in this book has been compiled from sources deemed reliable, and it is accurate to the best of the Author's knowledge; however, the Author cannot guarantee its accuracy and validity and cannot be held liable for any errors or omissions. Changes are periodically made to this book. You must consult your doctor or get professional medical advice before using any of the suggested remedies, techniques, or information in this book.

Upon using the information contained in this book, you agree to hold harmless the Author from and against any damages, costs, and expenses, including any legal fees potentially resulting from the application of any of the information provided by this guide. This disclaimer applies to any damages or injury caused by the use and application, whether directly or indirectly, of any advice or information presented, whether for breach of contract, tort, negligence, personal injury, criminal intent, or under any other cause of action.

You agree to accept all risks of using the information presented inside this book. You need to consult a professional medical practitioner in order to ensure you are both able and healthy enough to participate in this program.

# Table Of Contents

Chapter 1: The First Telescopes _____ 1

Chapter 2: Repairing The Telescope _____ 25

Chapter 3: A Mosaic Of The Crab Nebula, As Seen By The Hubble Telescope. _____ 61

Chapter 4: The New Universe _____ 88

Chapter 5: 1990 Introduction _____ 95

Chapter 6: Keenest Eye Space ___ 103

Chapter 7: Hubble Aperture Door Finally Unlocked _____ 120

Chapter 8: Five Key Discoveries __ 172

## Chapter 1: The First Telescopes

The magnificence of objects tiny and at great distances has prompted a study into Earth's natural environment, and the creation of itself. Within the field of astronomy as well as Cosmology, only four centuries have brought astronomers from small glass mirrors constructed by eyeglass artisans to sky-based telescopes "liberated"[1 from atmospheric distortion and background light from Earth.

Many landmarks began in the 17th century Europe which was a period when science was exploding and the people of the time must have been stunned by the invention of lenses. Space enthusiasts of both the 21st and 20th century should not be less enthralled by the incredible power of telescopes which are in operation today, in lower orbit about Earth. Early telescopes made space accessible to humans, while modern telescopes can peer back in time.

In 1608 the Dutch manufacturer of eyeglasses, Hans Lippershey, patented an extremely primitive telescope with the simple title of "Kijker," or "looker."[2Some say that Lippershey stole the design of another eyeglass retailer Jacob Metius, who applied for a patent only a couple of weeks later. Both were denied because of a variety of counterclaims as well as an opinion from the government that such a device could be easily copied. Jansen created the very first compound telescope and both he and Lippershey were awarded recognition for their work.

The Lippershey design was brought to the interest by Jacques Bovedere of Paris. He brought the design in the presence of Galileo Galilei of Florence, who designed his own telescope. it's Galileo who is the one who deserves the majority of the credit for the idea behind the telescope. In fairness this, the Galileo model improved the magnification power of previous attempts by over 20x. With this device Galileo was able to draw the moon's phases with great detail. He also observed the rings surrounding Saturn and could observe four moons of Jupiter. Galileo's invention was powerful enough to reveal the presence of a light-colored ribbon within the evening sky, which was later was identified by the name of Milky Way. Based on the observations he made, Galileo was convinced that the heliocentric model of Copernicus was correct and this belief led him to be put under house indefinitely until the year of his death. Galileo

Johannes Kepler contributed to the beginning of the 17th century through Kepler's Keplerian telescope. Kepler was a German mathematician, astronomer and mathematician, Kepler was also an astrologer before these fields parted with the advent of modern times. Kepler is considered "the pioneer of modern optics,"[3He designed the instrument with two convex lenses that increased the magnification, but also providing the image to the user upside down.

Christian Huygens, a Dutch scientist, mathematician and the founder in the field of quantum theory employed his own method to discover a new moon on the 16th of June in 1655. The instrument was made for studying the solar system and planets. Huygens introduced aerial and ocular (tubeless) telescopes as well as the first utilization of the micrometer. A 12-foot instrument was used to Huygens verified "Saturni Luna" which was also called Titan.

Sir Isaac Newton built the first reflecting telescope (as in contrast the term "refracting") around 1668. The basic system he built was known as the "Newton reflector" the concave primary mirror was placed against an elongated, diagonal second pane. It is the oldest functioning reflector telescope. When he was working on the spectrum of light, Newton employed what was described as a primitive spectrometer. However, the more modern instrument was developed about a century later through Gustav Kirchhoff and Robert Benson and their ability to divide light into individual wavelengths was a crucial component of the space exploration technology.

Newton

Laurent Cassegrain employed another simple style towards the end 17th century. named the "Cassegrain reflector" built on the folded double-mirror. Cassegrain was an ordained Catholic priest who had a constant fascination with astronomy and science and was a teacher in his final years of career.

A pioneering huge telescopes with a 40-foot length, was constructed around 1789, in 1789 by William Herschel in Britain. The biggest telescope in the 19th century was constructed within Wisconsin as the main focal point in the Yerkes Observatory.

The idea that a satellite telescope could be used was proposed during 1923 through German Rocket scientist Herman Oberth, who suggested that it could be an "space-bound" device in the book Die Rakete zu den Planetraumen. The 1930s were when Karl Guthe Jansky set his sights on solving the static issues in the telephone system . He worked as an engineer at Bell Telephone Laboratories and discovered radio waves in the Milky Way.

Sir Bernard Lovell conceived of an enormous radio telescope of 250 feet that could be directed at any location within the night sky. American Lyman Spitzer had proposed the idea in 1946 and he lobbied for thirty years to bring about the realization of the possibility of such a device. As a professor and researcher of Yale University, he continued to advocate for the benefits of a space-based system in comparison to ground-based models. In Princeton, Spitzer headed the National Academy of Science Ad Hoc Committee and also published the first research paper in response to a proposal that same year.

Spitzer

Lowell

In the years that Lovell was developing his invention the concept that a telescope located in orbit could work better than one located on Earth was being investigated. The 1940s were when astronomers as well as scientists believed that the atmosphere of Earth affected telescopes' ability to produce high-resolution images. This distortion is known by the term "seeing." The distortion is visible to those who stand outside in the dark and can see stars shimmering. The atmosphere of Earth is primarily a shield against ultraviolet and infrared light which are good for human life, but harmful to Astronomy. In the end, all the light produced by the sun and other living things can also impact astronomy performed on the ground.

The mid-20th century saw technology allowed us to develop radios and other devices that could detect electromagnetic waves throughout the electromagnetic spectrum. Instead of just observing the light spectrum, telescopes were created to also detect radio waves, analyze thermal and radiation with infrared light, and even detect gamma-rays and x-rays.

Making the Hubble Telescope

It was the dream of Lovell that first came true by constructing a telescope in 1957, however, thirty years of research had not yet resulted in the necessary delivery mechanism to create an orbiting instrument. The most effective Lovell could achieve could be one of the strongest terrestrial telescopes that was located in central England however, given that an instrument with all these capabilities would perform best in orbit, it became simply an issue of building an appropriate telescope and placing it in orbit. With the advancement of rocket technology in both the Soviet Union and United States were able to send satellites to orbit.

1957 was a year of great hope when it was the year that Soviet Union launched Sputnik, the first satellite to orbit. It was barely 193 pounds of ball that had a beeping mechanism Sputnik sparked the most intense period of the Space Race between the Cold War superpowers. It was the United States launched its first satellite into orbit three months later. NASA was founded on the 1st October of 1958.

First meeting of the Spitzer's committee was held in the year 1966 to conduct an examination of the potential uses for the telescope. It was in the year 1969 that the committee released Scientific Uses of the Large Space Telescope in which they urged for the development and construction of a prototype start. To make this happen the support of NASA was required. Wernher von Braun who was a ex- Nazi official who was the most prominent rocket scientist at NASA, was able to think about the possibilities and had suggested a tiny mirror with a diameter of 120 inches.

As of 1971 George Lowe, then acting administrator of NASA granted approval to the Large Space Telescope Steering Committee to conduct a series of studies on feasibility. The process of fundraising was set to begin. However, the cost was quite high at $400 million. This was an unpopular proposition for an economy that was recuperating of World War II and Korea and waging an "police actions" during Vietnam. The most recent Apollo space missions Moon have been awe-inspiring however, the cost was not. The request for funds was rejected from the House Sub-Committee in 1975, however, the subsequent collaboration in conjunction with the European Space Agency proved a diplomatic boon, decreasing the total cost. After a mirror reduction from the initial 3 meters down to 2.4 The cost of the project was reduced by half and the total in the range of $200 million was approved by Congress in 1977.

The first steps towards Hubble's construction was swift. The contract was awarded to Perkin-Elmer to build the optical assembly and mirror and grinding of the mirror's primary began in December 1978 in Danbury, Connecticut. Lockheed Missiles and Space Company is situated within Sunnyvale, California, was contracted to construct the spacecraft, as well as the support systems. The following year, the future issues with maintenance after the telescope was placed in orbit were expected, and training missions were launched for repairs, replacements and upgrades for all the components.

Computer-based lens-grinding made the mirror with a slickness that a mirror of the same size could be without flaws over six inches. However, when the release date was set for 1983, the launch was then scuttled by several reasons. The first was an unfinished optical assembly, even though it was completed of the mirror. As a result, the timeline was moved into 1984. This date was not achieved due to optical issues that took five more years to be resolved. Another launch was planned for 1986, however on the 28th of January the year that was when the Challenger spacecraft accident ended the project for a duration that lasted two years.

A backup mirror made by Kodak

Hubble Space Telescope Hubble Space Telescope would not be the only telescope based on space however, the limitations of similar endeavors were exposed by Hubble's capabilities. The telescope was named in honor of Edwin Hubble, a leading observational cosmologist in the 20th century, who virtually created this field, which was observational extragalactic. Similar to his nameake telescope, Hubble "changed the ways we think about the universe forever."[4 With having earned a Ph.D. on astronomy from Chicago University, he was given the opportunity by George Ellery Hale, founder of the Mount Wilson Observatory in Pasadena, California, to join the team. The invitation came while Hubble put the final elements of his thesis and was preparing for his oral exam. Hubble's telegram message to Hale was awe-inspiring: "Regret cannot accept your invitation. I am heading off to war."[5He joined the infantry, then returned back to America U.S. five years later and finally made it towards Mount Wilson.

Hubble

After returning, Hubble observed what he believed to be an nova star exploding within the Andromeda Nebula. Hubble later discovered that he had observed the "variable star,"[6which is officially known as the "Cepheid" which was able to gauge distance. At that point, Hubble began to study Nebulae. Within several decades, Hubble realized that the universe was expanding at an alarming rate. This led to what was later to be known as Hubble's Law. It's basically an assertion of "direct relationship between the distance of an object and its recessional speed as measured by red shift."[77 Hubble was a key figure in the design and development of the

Hale 200-inch telescope at Palomar Mountain, and he was the first person to make use of the telescope.

The European Space Agency contributed important technology to the long-running project, which included The Faint Object Camera, the two first solar wings which power the craft and a talented team of engineers and scientists in the Space Telescope Science Institute in Baltimore, Maryland. Since Hubble's launch in 2004, Europe is given 15 percent of the telescope's observations time. The Hubble Project's European Science Archive is located in the European Space Astronomy Centre [sic(ESAC). (ESAC) close to Madrid. Madrid. It was managed in the Space Telescope European Coordinating Facility of the European Southern Observatory near Munich. Its Space Telescope Science Institute selects the telescope's objectives and manages the data from Johns Hopkins University's Homewood Campus in Baltimore. The control of the aircraft is managed at the Goddard Space Flight Center in Maryland.

The argument for a space-based observatory has at present clear to those working in the industry of ground-based telescopes or anyone who is a pupil of the field of astronomy. The light that travels through the universe, as seen millions of years after it has traveled is altered, interfering with Earth's atmosphere like seeing objects through a glass of water. These distortions are visible when you observe what is known as the "twinkling" of stars. The space telescope remove distortions however, in Hubble's case the entire spectrum of view that spans infrared, visible and ultraviolet wavelengths is accessible.

On the 24th of April, 1990, the spacecraft Discovery was able to reach 600 km in altitude to ensure a safe orbit. This was a record for the shuttle. A famous NASA image depicts Hubble suspended above Discovery's cargo bay, which is 332 nautical miles higher than Earth. Hubble's orbit is circular, having a diameter of 557 km and incline to 28.5 degrees towards the Equator. The Canadian-designed Remote Manipulation System Arm (RMS) was in place throughout the pre-deployment process which also included the extension of the solar panel and antennas. The deployment began the next day.

The heart of the Hubble Space Telescope lies the principal mirror. It measures 2.4 metres in size and supplying light to five instruments in the spectrum of optics. Hubble is home to three cameras, including the Wide Field Camera 3, the Advanced Survey Camera, and the Near Infrared Camera. Two advanced spectrographs are two of them: the Imaging as well as the Cosmic origins spectrographs.

Its Wide Field and Planetary Camera is a versatile camera with the capability of capturing celestial objects across a broad spectrum of wavelengths and wide viewing distance. From ground-based amateur telescopes , to Hubble itself, Hubble itself, this type of technologies are the "easiest method to begin in taking pictures from the night sky."[88

In recent years, more advanced cameras are designed specifically for Hubble in a number of. For instance, the Wide Field Camera and Planetary Camera 3 (WFPC 3) dedicates one channel to visible and ultra-violet wavelengths. The other channel is reserved for the infrared camera. It covers all of the spectrum of optics. The detectors that are sensitive are solid-state identical to the ones found used in the majority of digital cameras. To detect the spectrum of the visible CCD cameras that are silicon-based or a video camera that incorporates an charged-couple device, or a transistorized light sensor in an integrated circuit is employed. CCD cameras are CCD camera functions as a camera that simply "manipulates the electrical signals into a specific kind of output."[9But in the case of Hubble, that's the point at which the resemblance ends. The WFCP3 has 16 megapixels and an extremely sensitive high-quality, low-noise array. It was developed in collaboration with both the Goddard Space Flight Center of NASA as well as Space Telescope and Science Institute in Baltimore. Space Telescope and Science Institute in Baltimore.

This Advanced Survey Camera within the second series is an alternative to an initial Faint Object Camera, installed in the third service mission. As accurate and faithful as the original it is the Advanced Survey Camera accesses a vast field of view almost twice as large as its predecessor, the Wide Field and Planetary Camera. The new camera can increase Hubble's possibilities for discovery by ten times. Hubble can capture large portions of sky in full detail while also performing spectroscopy using a specific instrument known as the "Grism"--a combination of grating and prism through that only the preferred wavelength can be passed.

The Advanced Survey Camera operates three sub-instruments. In addition to being the Wide Field Camera, which is able to detect distant objects and examine the way the universe has evolved and changed over time, the Survey Camera contains a high-resolution channel. In general, this channel is able to discern detailed light coming emanating from galaxies' central regions that have massive black holes exist. It can perform similar functions with normal galaxies, star clusters and gaseous nebulae, where unobserved planets can be hidden. The Hubble's contrast in bright objects is increased to a ratio of 10. Furthermore to this,"Solar Blind "Solar blind" is able to block light from visible sources so that you can observe faint ultraviolet radiation. Because of its flexibility High-Resolution is able to analyze weather patterns on other planets as well as auroras, like those observed on Jupiter.

This Near Infrared Camera is employed to discover distant galaxies that are not visible within the visible spectrum in the visible spectrum. The light emitted by distant galaxies is stretched from the visible to infrared because of expanding space and some was taken up by hydrogen intergalactic. Infrared capabilities have been vital for creating this Ultra Deep Field Survey, which is a composite of several observations made over the years that provide a detailed and narrow view of the universe. It is compared to "looking through the lens of a 2.5-meter straw."[10The Ultra Deep Field Survey is a unique study of the universe.

Together with together with Near Infrared Camera is the Multiple Object Spectrum. The light spectrum is between 800 and 2500 nanometers, and be able to penetrate dust clouds hidden by Nebulae. The unit that is combined is known as NICMOS.

The NICMOS operates in a cold and dark environment. Scientists must make sure that light from distant sources is recorded, and not the component's heat. Multi-object spectrometers and infrared instrumentation is maintained at -321 degrees Fahrenheit which is 77° Kelvin. In the early years it was located in an ice-cold cryogenic chamber that resembled a thermos. In the event that one expected modern technology to handle this cooling procedure, a traditional piece of nitrogen ice, weighing 300 pounds was employed instead. The cryo-cooler used for the operation was similar to the standard refrigerator you would find in your house. The majority of the work performed by NICMOS has been replaced by the 3rd Wide Field and Planetary Camera however the future of the first component is not clear, because the WFPC 3 isn't immune to malfunctions. NASA believes it will eventually be able to reach galaxies that created less than 400 million years following it was formed following the Big Bang.

If these amazing feats of optical magic are to be successful, a mechanism must be in place to keep the entire telescope and telescope in a completely still position to photograph the distances in long exposures. A different set of instruments, such as those of that of the Advanced Survey Camera, can serve the essential functions as well as analyze and observe scientifically simultaneously. What are known as the "Fine guidance sensors" offer the general craft the stability it requires while also providing information for locating distant celestial objects as well as measurements of the allied distance and time. These "Fine Guidance Sensors" are composed of a huge structure housing a range of lenses, mirrors beam-splinters, servos, prisms and photomultiplier instruments.

Sensors are used to move the telescope in order to measure celestial phenomena. Two sensors are dedicated to directing the lens towards its goal, while the third sensor is used to make a myriad of observations. One could only imagine the challenges in determining a steady position in a floating setting in particular when the camera is roughly as big as a bus, with its camera is the size of an infant grand piano. Hubble is able to move effortlessly along its optical axis but not in a complete end-to-end manner however, it is an axis that is a log roll. Its movement is limited due to the requirement to keep sunlight shining onto its solar panels.

Sensors sense the slightest movement of the device even in small amounts, and instantly restore the initial location. This is so effective are the interconnected systems that the Hubble's stabilization has been compared to keeping the laser beam "on the size of a dime, and at 400 miles for 24 hours."[1111

In an academic setting sensors can look for wobbles that indicate that an object is orbiting. It can detect double stars, measure the diameter of a star's angular and also refine the position and brightness scales on any object in the universe. As for measurement methods, it is able to determine the actual distance scale of the universe.

With its ability to separate light from the background, the spectrograph is able to reveal its chemical makeup, temperature and movements of various comets, planets, stars interstellar gas, and galaxies. Through a unique method the light is absorbed by an slit that is long, and before it reaches the grating the spectrograph is simultaneously recorded across every one of the 1.023-pixel rows of the detector. If the slit is directed to the direction of the nucleus in an object, Hubble can measure how the speed at which a galaxy is rotating across a variety of distances from its central. In the center of galaxies, where huge black holes are located The spectrograph has discovered and determined the mass of a few dozen.

The STIS was used for the continuous study of dust and gas blowing by massive binary stars with unstable properties like Etz Carinae. They are located within our own stellar region around 8,000 light years away. In the case of the planet HD 1897 the STIS was able of determining the whole body's blue color through the spectrum of visible light. The spectra from the transiting star HD 209458 produced luminescence curves that were so precise that the instrument could determine what amount of light that was absorbed by the atmosphere of the planet and for the very first time ever, the planet's atmosphere was recognized as comprised of oxygen, hydrogen and sodium.

The STIS enhances that of the Cosmic Origin Spectrograph. It is an instrument for high sensitivity medium - and low-spectroscopy on objects of the astronomical world in between 815 and 3200 A. Its Cosmic Origin Spectrograph improves Hubble's capability to search for and capture faint sources of light that are difficult to reach from ground. It was designed to study massive structures within the universe, and focuses on the development of galaxies, the birth of planetary and stellar systems, as well as the interstellar cold medium. While STIS spectrograph is an STIS spectrum is an "all-purpose"[12component capable of handling large bright objects and efficiently, it is a more specific instrument. Cosmic Origin Spectrograph (COS) detects incredibly faint levels of ultraviolet light coming from cosmic sources like quarks.

Many have written about the amazing degree of electronics found in Hubble's many components, all contained in a protective shell that is covered in blanketing material to guard against the most dreadful threats, even when in lower Earth orbit. Hubble's New Outer Blanket Layers, known as NOBLs are said to were discovered to be "taken the brunt"[13of the most severe effects of space, which include extreme temperature fluctuations solar radiation, extreme temperature swings, and micro-meteoroids dissolving the blanket material. Fortunately, the majority of the wear in the initial years was cosmetic. The initial service mission discovered the exterior in perfect condition, however on the second inspection, over 100 cracks of more than five inches long had been discovered however no one knew what time the next visit would be scheduled. To shield the exterior from constant bombardment the four patches of aluminized single-sheet Teflon were applied , and remain in the same place. When it comes to taking care of the shell astronauts have learned many things about space degrading along with the overall rate of degradation which can be used in other missions. One of these issues is the possibility of internal dangerwhenever there is active electronics and heat production and rising temperatures inside. Certain blankets allow for internal cooling with an integrated radiator.

A variety of intricate control of altitude enhances stabilization of this instrument as it performs its duties. Concerning the maneuverability of the craft the thrusters are completely absent. Concern about contamination of the observations is a significant concern and so is Hubble's overall life span. Reaction wheels that are heat-free move the Hubble to its desired position, and the gyroscopes track its movements for the accuracy. The wheels depend on the "elegant" principle of Newton's third law; thus, "if one of the wheels turns clockwise, Hubble will turn counterclockwise."[14] Finally, the "Fine Guidance" sensors lock onto "guide stars" for accuracy.

## Chapter 2: Repairing The Telescope

On the 20th of May, 1990, just one month since Hubble's first deployment the mission was hit with its first-ever unexpected glitch. The first target was set in fact, Hubble was producing data and images, however, it was "slightly misshapen,"[15 and a blurry, unfocused distortion was created in the images. When it was time for the launch, no one knew that the mission was in danger.

Astronomers who worked within the visible spectrum eagerly looked forward to images and data that had improved clarity from outside of the atmosphere. The spacecraft had spent 15 years creating Hubble's Wide Field and Planetary Cameras and scientists and people of all ages were thrilled by the news, eager to find out whether Hubble could be saved. NASA itself had invested more than $15 billion , and was likely to be pressed to get over such a costly loss to their telescope programs.

The fault was discovered in the fine grind of the main mirror and, as it turned out that the error was committed in the year 1981, which was nine years prior to the launch. In essence, a portion of the telescope had been close to being sighted. Repairing the telescope was not difficult, but the procedure was a challenge and pushed the limits of precision. "The answer for the telescope with a nearsighted lens is similar to that of the nearsighted person,"[16said Kiona Smith from Forbes and that's corrective lenses. Through a complex and intricate procedure five mirrors that could be adjusted were created to redirect light from the main mirror prior to it reaching the telescope's instruments of science. Within the halls of this agency, the most hated terms"spherical aberration "spherical aberration" were often used.

NASA was clearly in trouble, and a lot depended on an effective repair. Hubble's press coverage was not a good one in the days prior to debut, yet it become worse in the following three years. Much was at stake by the Hubble project that similar programs were dropped in a frenzied manner, and accusations of incompetence were made in the press. The New York Times asserted that had the mirror contract been given to Kodak-Iter the issue would have been identified in time. This Space Shuttle Program itself survived Congress with just one vote while that's when the "Moon to Mars" project was shut down.

Astronauts tasked with fixing the telescope were well aware of the work's essential nature. If they did not succeed, the it was likely that the political will behind Hubble would be lost. Hubble Program would die out. In the Mars Observer mission failed short of reaching the red planet and there were rumors that Hubble's WFPC 2 might be faulty. President Clinton privately advised NASA that any failures would result in the agency being restructuring. From the space shuttle to the robotic arms, the astronauts were expected to be able to work "with the harmony of the ballet."[17The president was adamant that the agency would be restructured.

What Smith described as the "contraption"[18was officially referred to as COSTAR. Corrective Optics Space Telescope Axial Replacement COSTAR. COSTAR. The way the media described as Hubble's "spectacles"[19was accompanied by an estimated price of $50,000,000. Everyone was eager to climb but the chance to lead was presented by astronauts Story Musgrave who had been through numerous spacewalks and missions. A natural choice, his hard-earned training led colleagues to laugh about how the most peace found was in the chair of a dentist.

The mirrors of COSTAR, each about the size of a nickel required to be perfected in three years, and so did the new camera to compensate for the mistake. The Hubble's fix earned the image of a daring show, and at that time it appeared that nobody believed it could be done. the agency was in too much "Hubble trouble."[2020 NASA became a target for late-night comedians, and previously anonymous camera engineers sat in a fishbowl that was open to the public. For some, the notion of causing the camera to be out-of-focus similar to the mirror appeared ridiculous.

NASA required three years to create the correction. When the spacecraft Atlantis was roaring towards the atmosphere's upper levels with the new cameras, scientists shook their heads at the sight of the new camera rattle and shake throughout. Installation took about 35 hours of spacewalks that began on the 3rd of December 1993. The astronauts were instructed to repair everything that required fixing because the cost of each spacewalk was so expensive. Two rate-sensor units were upgraded after a quarter of the gyroscopes were damaged. The solar arrays followed. They rolled smooth into the storage canister but the second froze due to a bent strut that was then cast with a reluctant ease overboard. It was the Wide Field and the Planetary cameras were both upgraded in the hopes that they could aid in the reduction of blur.

It was the WFPC 2 was previously an underground unit, but it seemed to change as planned and an optical correction was added. The fifth day COSTAR was added to replace the overheated solar array drive electronic components. A few weeks were needed for the alignment of the telescope to its new components.

Astronauts didn't have a way of being sure if the new pictures were a significant improvement. At the surface, champagne poured and tears gushed when the first pictures arrived with stunning quality. There would be more snags in the following 15 years however, the designers and teams grew more adept at dealing with every new challenge. Despite the obvious defect at the heart of the mission, Hubble will continue to produce a number of remarkable achievements despite the fact that its mirror was temporarily damaged.

Producing Results

In spite of all the challenges that were involved in the design of the scope, that included the contracting of several organizations, using the telescope is a breeze. The Hubble telescope is operated by its own operation centre, called it's called the Space Telescope Science Institute (STScI) that manages the telescope on its own. The STScI with an employee base of hundreds of engineers, scientists and astronomers, runs the telescope mostly for NASA and occasionally the European Space Agency, but it also runs this telescope to amateur astronomers as well.

The telescope was planned to be launched in the beginning of the 1980s, NASA had already formulated specific plans for the telescope a few years prior to when it was launched. The projects were immediately prioritised and took a considerable amount of time to finish.

The first project was the creation of the Hubble Constant, discussed further in Chapter 5.

Every year each year, the STScI will create an annual plan for the use of the telescope. This is achieved through receiving and reviewing requests on where to focus the telescope and where to focus it on. The majority of requests originate directly from NASA and the European Space Agency, and the STScI collaborates with both agencies to come up with concepts for the use of the telescope. Individual projects, that can be suggested by any person in society, is included.

After the STScI is aware of what Hubble will capture pictures of throughout the duration of the year, it employs software to create a feasible schedule for all the projects scheduled to take place. The STScI also has additional software that controls and monitors every instrument of the telescope and give it control over the telescope's movements. Operation of the telescope can be synced to the timetable, however in the event of an unpredictability event taking place the telescope could be altered quickly, which happened in 2009, when an unrecorded and untracked asteroid hit Jupiter.

The HST is connected with the Tracking and Data Relay Satellite System which is a mix of satellites on orbit as well as communications links that are located on the ground. Alongside the system being used to control the telescope, the STScI depends on the system to collect images from the telescope.

If the images of the telescope are transmitted to the system, data is sent to both STScI as well as NASA. STScI transmits the data to algorithms software which analyzes the data and then stores the data.

The most interesting aspect this telescope has is the fact that it can be used by any person in the general population can make use of it. Apart from coordinating public outreach programs The STScI is able to offer a method that permits amateur astronomers submit proposals to request that the telescope is used to carry out a project that they would like to use it for.

It's certainly not as easy like asking STScI to examine an object. The proposal will be subject to peer review by the STScI and the person submitting it must have an concept of what they're planning to do. The peer review will determine if the proposal is an appropriate application of the telescope's time however, during any one year the STScI will take approximately 100 proposals.

When a proposal from an applicant is chosen When a proposal is selected, the STScI helps them by providing the software to let the applicant choose where the telescope should be pointed for the work. After the applicant has figured it out using the special program of the STScI they can coordinate all the details with the STScI and it will manage the telescope and carry out the task.

The telescope's first focus was the brilliant Star Cluster 3532, 1,300 light years away, finally was given the attention it deserved. It is part of the constellation known as Carina and is the vessel's keel. Argo. It is also known as The Wishing Well Cluster, it is believed that it resembles silver coin that are dropped into the well. However, some have called this the Football Cluster, which can be observed by anyone with a clear eyes in the southern part of the hemisphere. The cluster is around 300 million years old however, iconic images of the nebulae aren't likely to be the result of a Hubble shotdue to the difficulties it faced in its early days. The most popular image was captured using The Wide Field instrument of the Chilean the La Silla Observatory about 23 years on.

In August of 1990, Hubble came within a 0.1 second of studying the ring of material that formed around what remains of Supernova 1987A. In October, the first scientific study on Hubble was published by Tod Lawes, a researcher at the National Optical Astronomy Observatory in Tucson, Arizona. The focus of the paper was on the physical environment of black holes which is still a concept which was developed by Einstein and other scientists.

In the second week of the year 1991 Hubble precisely measured distance to a nearby galaxy and the telescope offered an added benefit by studying in the Large Magellanic Cloud which is an asteroid from the Milky Way and the supernova's home. The position of the telescope's viewing point of the Earth was 169,000 light years away from the star. Then, two months later, Hubble's initial photographs of Jupiter were made public, including"the "Giant red spot." The view of

Jovian weather the red dot is believed to be one of several large hurricanes. ammonia crystals from warmer gases are transported from deep within the atmosphere to the cloud's upper layer. In further research, it appears that the shape oval of the spot appears to have changed into an elongated shape.

The capacity of a telescope to identify ancient objects was proven in the month of January, 1992, when astronomers spotted an element that is rare, boron within an old star. This was a indicator of the beginning of the universe. The star of 7th magnitude is just 100 light-years away. Boron is generated through high-speed and powerful cosmic rays from supernovae. This "boron" Star could also be "fossil" for the events that led to the formation of the Milky Way. Its discovery could trigger some adjustments of The Big Bang Theory, appearing only in the ultraviolet spectrum. In the beginning, Big Bang may have already produced some of the structures in the first minute, including beryllium and boron.

In search of black holes that are tangible physical entities Hubble could find fuel being pulled from the abyss's gravitational core of galaxy NG4261 on the night of November 19th 1992. The galaxy's giant elliptical was 100 million light-years away. A dust spiral that is 800 light years away from Earth powers the black hole to the constellation Virgo. The speed of the swirling gas suggests that it is 1.2 billion times as massive as the Sun.

On the 8th of June in 1993, the brand new telescope tackled one the most difficult issues to be asked in circles of astronomy: the size and age of our universe. The result of the search was what's known as"the" Hubble Constant, a unit of measurement that is used to explain the expanding universe. The first rate was set at 160 km per second for every million light years. The rate is subject to the caveat of 10% uncertainty. In this particular observation, Hubble targeted two "variable stars." Two fields were examined by exposing them every 22 minutes for 14 months.

Scientists believed for a long time that the expanding universe was slowing in its progress and that the speed created due to the Big Bang had lost much of its energy. However, with the study of supernovae, stars which self-destruct during their last cycle of existence, Hubble confirmed that the contrary is the case. It is believed that the universe expands at a faster rate , just as it has done since thousands of years. This realization suggests an existence "dark material." While we may not comprehend about this intangible force, we now observe that it is greater than the gravity force.

The vastness of the universe are in the minds of any scientist or student who has thought about the meaning of creation and wondered if the universe is infinity. If not then is there a clear boundary where everything of what is known, or that can be observed is over? The desire to explore further by trying to guess what's beyond the boundaries is intoxicating. Imagine the entire thing as something that connects to itself, creating an enclosed infinite, has been thought of as.

The questions are not valid because astronomers have been provided with only the "observable universe" as a model which is best found in Hubble's archives. The telescope's capability to offer better photographs of the distant galaxies as well as other deep space objects was discovered shortly when the telescope had been set in

1993. Over a 10-day period in November 1995, it directed towards a region of the Ursa Major constellation. Ursa Major, using a combination of long exposure times and short exposure times in order to take a picture of objects deep in space.

The Hubble Deep Field project, later to be known as Hubble Deep Field, provided an image that was clear enough for scientists to identify thousands of galaxies made up of various shapes. Furthermore, some galaxies seemed to be merging with one another. Based on other observations taken from other telescopes, it was discovered that galaxies seen in the image were twelve billion light-years away, which suggests that stars were beginning to form within the first billion years of when the Big Bang.

What is it that made the Hubble Deep Field so stunning is that it encompassed approximately 0.000002 percentage of our sky or about 1/500,000th. This means that the image suggested that the universe was composed of millions of galaxies. The telescope was utilized to create a similar picture called Hubble Deep Field South, that showed similar results in another region in the night sky.

The telescope was employed to create similar images, which could take us deeper into space, such as Hubble Ultra Deep Field. Hubble Ultra Deep Field which was captured between September 2003 and January 2004. This image featured galaxies which were formed within a few hundred million years following being formed after the Big Bang, sharply reducing the time that it took for stars to form following that Big Bang. The image also managed to record and photograph the individual stars that are 60,000 light years from each other.

Alongside giving Astronomers a glimpse of the size that the universe is, they also showed that the cosmic scale could be "homogenous," meaning that the same physical principles were in use everywhere. This confirmed the "cosmological principle" that has been held for a long time by astronomers who believed that they believed that the Milky Way galaxy's position in the universe isn't exclusive or unique in any way.

The Hubble model of the universe consists of 93.3 billion light-years in size. There is no doubt that a larger "universe" exists and scientists aren't capable of determining the possibilities. Utilizing methods such as the Bayesian Theory, a mathematical formula for probabilities which is based on the force of an expected result is pushed against the unknowable by the force of existing evidence. When determining the size of the universe in the universe, the Bayesian method estimates the universe to be 250 times larger than is Hubble's 93.3 billion light years, which is a 7 trillion-light years across.

The universe is vast however, with billions of massive galaxies, they can meet each and each other. Hubble telescope Hubble telescope has taken pictures of galaxies "colliding" with one another. The collision is caused by the gravitational pull exerted by the galaxies against one another and could result in merging of the two galaxies, or in the "cannibalization" of smaller galaxies by the larger one.

Hubble has recovered its images and information based not just on distance, but also on time. The current measurements are converted to "the light year," which is the distance that an object travels over one year, at speeds of light. The distance from "here to where" is inextricably linked to "then to the present." The light Hubble sees is like it did in past times dependent on the distance the telescope is observing. The HTS has observed galaxies that are which are so small and distant that the pictures recall the time of the Big Bang. New discoveries have set our observational clocks back an age that was only hundreds of millions of years after the famous event. Based on Massimo Stiavelli, a scientist at the Space Telescope Science Institute, "Hubble brings us within a stone's reach of that Big Bang itself."[21]

The galaxies of the early days show the chaos that occurred in the initial stages of creation, as opposed to spirals that are so common to our modern-day. Each shape is apparent like that of the "toothpick" as well as the "bracelet" arrangement. In the beginning, Hubble receives about one photon per minute. This is only a fraction of what we receive from later galaxies, where we receive around 1,000,000 each minute. In the search for the beginning of the observable period one of the smallest early galaxies are visible to us with brightness levels one-tenth of one billion lower than the human eye could discern. The stars that are visible using the Hubble's top optics are those that are described as being"the "second generations," after the first instances of matter produced by the universe's new creation. This is possible thanks to an instrument that is no bigger than a phone booth recording light emanating from the central point of the universe, long before the Earth was born.

Similar to this, the inhabitants of Earth have believed for a long time that our solar system originated as an elongated disc. Hubble shows how widespread this pattern is in the birth of stars and the bodies they surround. In the Orion Nebula, about half the stars in the early stages are covered by dust and gas structures in the beginning stages with a large proportion of them discs.

With this long history of investigating the issue we have a feeling that the universe is speeding up through the unexplored phenomena that is dark matter. Astronomers are constantly releasing new information on the way supernovae have changed in the course of time. The farther away the stars that explode are, the less bright they are and the more time-stretched (red-shifted). The universe is accelerating according to scientists like Neal deGrasse Tyson, represents the possible closing of the universe where all atomic as well as molecular material, including biological or inert, is released and disappear.

In the middle of January 1994 Hubble made public data about its findings of Eta Carinae an enormous unstable star. It is prone to violent eruptions "Carina" can be seen as 4 million times more bright than Sun and is 150x larger. In the past century it was among the most brilliant celestial objects even though it was a long way away. What Hubble saw was the last moment from the star that was dying seven hundred light years away being stripped of its primary elements in a massive cloud of charged particles within the constellation Carina. The loss of mass is equivalent to one Sun every thousand years and the probable result is that the star will explode within one million years.

Four months after Black holes became the center of attention when observations focused on the elliptical galaxy M87, which is 50 million light years away. A supermassive black holes was found in the galactic center as it was a "gravitationally collapsed object that exhibits rapid rotation around its core."[22This was a remarkably definitive evidence for previously proposed entities that were predicted over 80 years ago according to Einstein's Theory of Relativity. The matter inside weighs more than 3 billion Suns packed in a space that is no bigger that the Solar System. Over the years, reports of flaring were recorded as sudden increases in brightness, as the phenomena interacts with surrounding materials.

The world is abuzz with memories of the comet that hit Jupiter and was covered on the news at night in 1994. Numerous pieces of comet Shoemaker Levy 9 in a series of impacts on Jupiter's atmosphere for a number of several days. The first documented instance of two collisions between astronomical objects with each other, it was reported that the Jovian air was observed as "bruised" due to the impact of the comet.

The discovery was made through Carolyn as well as Gene Shoemaker and David Levy The Shoemaker-Levy pair had probably been orbiting Jupiter for more than 10 years before it was caught and broken by the gravitational force of Jupiter. Each fragment made a spectacular collision with the force of 300,000,000 Atomic bombs. Gas plumes were emitted over 300,000 km and the temperature to 71,000 degrees F.

Hubble telescope images of the impact

A view of the impacts sites on Jupiter

In November Hubble began to study the features on Saturn's surface moon Titan which is a subject that is of great curiosity. It is larger than Mercury however smaller than Mars Titan's atmospheric structure is 4 times thicker than Earth and nitrogen is its main ingredient. The moon's hazy appearance was measured with bright and dark is a part of a 16-day rotation with one particularly large area that was larger than Australia. Titan was at the time the only solar system other than the Earth with an ocean confirmed. However, they are mostly made up of ethane, and methane. The surface of the liquid is harder to freeze than granite. Hubble utilized their Wide Field Planetary Camera 2 (WFPC 2) in the infrared spectrum, resulting in 14 "noise-free"[23images that were free from interference caused by signals. All in all 50 images were created and the information will be of vital significance in scientists on the Cassini Mission as it entered the Saturn region.

Hubble was back on Jupiter in February 1995, discovering an oxygen-rich surface on Europa which made the moon only the third spacecraft of the solar system that has some. However, Europa holds a "tenuous atmosphere"[24In which the surface pressure is one hundred billionth the pressure of Earth. If all oxygen was trapped on the surface the area would be more than 12 Astrodomes in Houston. Hubble's ability to discern this small amount of oxygen is amazing, however, care is required to ensure that you don't misinterpret the presence of oxygen as the sign of existence. November was filled with star births within the Eagle Nebulae, nicknamed the Pillars of Creation. It is located in the constellation of Serpens It was first identified in the 17th century by Swiss scientist Phillipe Loys de Cheveaux in the mid-18th century . It was thoroughly observed through the eyes of Messier who discovered it in 1764 who named it M16. "Eagle" is approximately 70 light-years by 55 light years, and includes several regions suitable for the formation of stars. In the visible cloud that contain hydrogen, gravitation pulls the clouds of gas to collide to the side. If there is enough gas then nuclear fusion can ignite the central region. The information on the formation of stars is vital to our understanding of the origins of the universe. The nebula can be observed with low-power telescopes and binoculars and the famous Hubble images have inspired a lot of amateur science.

Images of composites of previously unexplored areas in the Universe were revealed on the 15th of January of 1996. It was dubbed the "Deep Field" wherein three42 minute-long sections of sky captured of WFPC 2 over ten days. The composite images depict an array of 1,500 galaxies at different levels of development from the tiny "keyhole" which extends into the past, all in the space of a dime located at 75 feet. The majority of the exposures taken in prior to release showed galaxies that were four billion times fainter than what our eyes can see. These distant galaxies were never observed by ground telescopes. One scientist identified the composite images as "deep core samples,"[25the equivalent of the astronomical version that of Dead Sea Scrolls.

The next day new data revealed evidence of a satellite orbiting the Star Beta Pictoris. The mysterious object was hidden by an outer region of dust that was changed shape by the gravity pull of the planet.

In the beginning of March, 1996 studies of the dwarf planet Pluto revealed basins and impact craters that were visible on the surface. They were discovered by studying the brightness of blue light. The majority of the surface of Pluto was photographed during its rotation of 6.4 days. It was revealed to be an intricate object. The images were captured using the ESA's Faint Object Camera. Pluto's size equals about two-thirds the size of the moon of Earth, and is only 12,000 miles from the sun. One scientist compared the images of the tiny sphere to "read[ingthe tiny print on a golfball thirty-three miles away."[2626

This Hubble Space Telescope took its 100,000th image on June 22 in 1996. The object was a quasar which lies 9 billion light-years away from Earth. Prior to the invention for the Hubble telescope scientists were confused about how to define "quasi-stellar radio sources" more commonly referred to as quasars. These were extremely bright objects that were that were billions miles away. They were emitting massive amounts of energy throughout the electromagnetic spectrum. They were brighter and more efficient than galaxies. Astronomers could not identify the size or the nature of the Quasars.

Astronomers have only recently begun to determine what quarks are following the Hubble telescope located their location in the galaxies' central regions. The result was that a quasar is the nucleus in an active galaxy, with a massive black hole that is drawing in and devouring stars close to it. As the stars move closer, a portion of the material is sucked away and it begins to form the "accretion disc" at the center of galaxy. The black holes do not emit light however the accretion disk is beginning to emit massive quantities of energy and light as they get closer towards the black holes. They eventually get heated to temperatures of million of degree Fahrenheit. The light produced by this small region can surpass the light of hundreds of galaxies that contain thousands of stars. Despite the fact that the Milky Way is 100,000 light years across, it's not a big enough galaxy to contain a supermassive black hole that could create one of these quassiars.

It also educated astronomers more about gamma-ray blasts. Astronomers and scientists have been able to recognize and identify gamma-rays, which are the most radioactive radiation in the electromagnetic spectrum for more than a century, but when NASA began to put satellites into orbit, they started to detect flashes of gamma radiation without having any clue as to where they came from.

The knowledge of astronomers of the gamma-ray bursts was only discovered in the late 1990s due to the Hubble telescope's capability to capture images of gamma-ray explosions combination with detectors for gamma-rays such as that of Compton Gamma-Ray Observatory. Although they are typically thousands miles away from us, gamma-ray explosions have such a bright brightness that one of them in 2008 could be seen by anyone with a naked-eye. However they are only lasting a few seconds up to March 2011, and, until that point the longest-lasting bursts that were observed had been just one or two hours.

The long gamma-ray bursts appear to be the final stages that massive star collapsing into black hole or supernovae. They emit more radiation in just a couple of seconds than the Sun can produce during its entire lifetime. In addition, the bursts appear to be narrow beams, moving at a velocity of light. Astronomers haven't yet been successful in determining the cause of the small gamma-ray bursts.

In March of 2011 The Hubble telescope, along with other gamma-ray detectors and telescopes witnessed the longest-lasting Gamma-ray burst that has been observed to date. The burst continued to emit bright emission for longer than a week after being observed, and after the HST was able to pinpoint the exact location, it was the galaxy was 3.8 billion light years distant, with the light emanating from the galaxy's central. Astronomers believe that this burst is the result of the massive star being near enough to the black hole supermassive in the middle of the galaxy. This started to strip it of material. The intensity and duration of the burst suggest that the gamma-ray explosion was generated towards Earth.

The sightings of gamma-ray explosions are extremely rare, in part because they only appear when the emission is reflected in specific directions. From Earth there are a lot of gamma-ray blasts that cannot be detected this is a great factor since should Earth was to be in the path of a massive radiation gamma-ray, even thousands of light miles away, it would ruin the entire atmosphere instantly, stripping nearly 50% of the planet's Ozone layer. However, the depletion in the Ozone layer will not be a problem due to the massive radiation produced by these gamma-ray blasts could nearly instantly kill all living things on earth, because gamma-rays travel at speeds of light there is no way to tell if it's happening before time.

The Hubble telescope's first image of the star's surface was of Betelgeuse, a supergiant red within the Orion constellation that is nearing the end of its lifespan. Betelgeuse has a radius that is 600 million miles. which is 1,400 times larger than the Sun and billion times its size. The photo that was taken from 650 light years away in the ultraviolet wavelength included a mysterious hotspot at the surface. The temperatures in this region exceed more than 2,000 Kelvin higher than other locations.

The second service mission for the telescope scheduled for February 11-21 in 1996. The most significant upgrades included installing two technology-driven instruments that were a combination of the Near Infrared Camera and Multi-Object Spectrometer (NICMOS) which enabled Hubble to observe the universe's infrared spectrum also the Space Telescope Imaging Spectrograph (STIS) which allowed celestial objects to greater clarity and was developed for the purpose of searching for black holes. The two instruments have been calibrated to the flawed primary mirror substituting the Goddard High Resolution Sptrograph as well as the Faint Object Thermograph. This technology was not in use in the beginning during the missions.

The upgrade proved to be worth it in May 12th, 1997 when recording the first spectrographic fingerprint of a black-hole. Hubble captured the gas's motions captured by the gravitational pull in the central region of the M84 Galaxy and the gas in the grip of the black hole was measured at speeds as high as 8,000 miles an hour.

A plume of dust and gas was seen from an eruption of volcanic origin on the moon's outermost moon, Io, at a distance of 2,500 miles while it was aiming at Jupiter along with its moons a little at least a month later. The eruption began at a speed of 2,000 miles per hour, however on Io the plume climbed to more than on Earth due to the fact that the atmosphere is extremely thin and gravity is weak, allowing for very little resistance.

Of all the research Astronomers could conduct by looking outside their planet into space, discovering the mechanism that causes the universe to move was the most important. The first concrete evidence of the possibility of an expanding universe from five years prior was confirmed by the Hubble telescope in September 1998. With the incredible precision that was used in these missions as well as the Hubble data from its couple of years in orbit, many views of space were shattered through this one single observation.

A month after, NASA assembled a "host mission" with the intention of verifying the effectiveness of the new equipment during the next service mission within the operational environment. The crew was transported via shuttle Discovery spacecraft to the telescope and was accompanied by an old friend on board. The mission's leader included John Glenn, the third man to orbit the Earth as well as the very one of the first American spacecraft to circle the Earth.

The Ring Nebula was discovered 200 years ago by French astronomer Charles Messier, who cataloged it as Messier 57 or M57. On the 6th of January in 1999, Hubble provided a view of the ring, of which Messier would have only dreamed. Hubble observes a cloud of haze, which is the glowing remnants of the sun-like star. The thing that makes this ring unique is the fact that it is filled with matter. The nebula measures just only one light-year across and the ring is the shape of a football which has run out of hydrogen resources. The glow emanates from the interior radiation that interacts with a massive mass of Helium. After about 10,000 years the image will get less brighter and blend with the surroundings because its companion star isn't strong enough to create an explosive final.

Hubble went into safe mode in an automatic reaction after the failure of four of six gyroscopes in November 13th 1999. Hubble maintained its safety by pointing its telescope towards the Sun so that the solar array could be powered and the antennae would work to communicate with Earth. All research was stopped to protect the remaining gyros and the mission overall was not at risk. Six of them are older and were known to have problems over the course of 50,000 hours use however, the more modern ones were more secure. After the system was powered up again and there was a strange sound coming from the electrical system. After jogging again it was discovered that the gyros had misread the rate of change, appearing to indicate that something happened, when it wasn't. In this condition, close and fast-moving targets were hard to identify, however on December 19 it was time for the following service to be scheduled which included the gyros as well as the new computer were put in place.

A bizarre announcement was made in May 3rd, 2000 declaring that Hubble discovered the universe's hydrogen that was missing. The element's mass were created during in the Big Bang shortly after which it went extinct. Hydrogen is likely to have been present in all galaxies that followed. However, through the light of quasars that pass through intertwined clouds of gas during the course of transit, hydrogen was discovered in invisible filaments that weaved their way through galaxies all over the universe.

Astronomers have been searching for the first hydrogen reserves that account for over fifty percent of "normal" matter in the universe. The remainder is locked within galaxies in an obscure shape. Hubble was at first in awe of it within the massive clouds that were in a hot, rarified state. Supercomputer models suggest gas filaments that are intricately interwoven in which hydrogen is concentrated in huge chains, with galaxy clusters appearing at the point where chains meet.

The birthing chambers of planets are plentiful and are, however, typically, they're hidden in clouded, hazy curtains of gas and dust. On April 26th 2001 Hubble observed the world's first tangible evidence of the growth of planets. Planetary "building blocks"[27 are located within the dusty disks of stars that are young in the Orion Nebula where they are "blowtorched"[28 by ultraviolet radiation from the brightest star in the region, making planetary formation difficult. Planets that are young have to "beat time" because dust particles remain together as stars attempt to break them up. The process has been compared to "building the tallest building in an tornado."[2929

Finding the first elements of the atmosphere of an exoplanet occurred in a sun-like body orbiting Star, HD 209458, 150 light years away. Seventh magnitude stars, such as sodium, was found on some of the planets. Hubble was never tuned to search for gases that one would expect to find in an atmosphere that is life-sustaining this discovery could be the first proof of existence beyond Earth in gas produced through living creatures.

In the Servicing Mission from March 1-12 2002 The HTS did not represent the only unit to have issues. The cooling system on the Columbia's cooling system Columbia failed. A flight rule that stipulates that an aircraft with just one cooling system is required to go back to Earth was twisted and the astronauts were able to reach Hubble for a near two-week time-span in that they were able to use the Hubble Space Telescope upgraded to the 21st century using the new Advanced Camera to conduct Surveys. The solar arrays that were flexible were replaced, which brought the capacity to 30% higher the power source. Power control units were upgraded and an entirely new cooling system was installed to cool the Near Infrared Camera that had depleted its cooling source that was a 230-pound chunk of nitrogen frozen. Spacewalkers were replaced with one of four reactions wheel assembly. With the first technological advancements over the past decade, data processing was has increased by more than 10 times. The first images taken by the Advanced Survey Camera were revealed on April 30th 2002. The images were not anticipated to be perfect, but were valuable in focussing the camera. The exposures were a bit fuzzy due to the atmospheric "smear," but they were much clearer than the ground-based shots of the same object.

The Crab Nebula Hubble recorded the dynamic of the "tremendous stellar explosion"[30that saw Matter and Antimatter driven at a speed that was close to light. The amazing view of the Crab Pulsar and the huge Nebula it powers, shows remnants of a supernova that occurred earlier in the year 900. The nebula spans 10 light years across and 7,000 light-years away distant from Earth located in Taurus. The exposure was made with the WFPC 2 with an astronomical wavelength of 555 nanometers.

## Chapter 3: A Mosaic Of The Crab Nebula, As Seen By The Hubble Telescope.

Columbia was lost. Columbia went missing in a devastating failure to reenter on the 19th of September 2002 following a 15- day mission. The service mission that followed was cancelled. In the meantime, Hubble maintenance schedule was stalled while officials and the federal government were examining the future of the shuttle program.

In the month of March of the following the year Hubble discovered an object that was evaporating. The "hot Jupiter"[31gigantic gaseous world HD 209458b was closely orbiting in the direction of its mother star and the atmosphere bled an alarming rate to space, with the majority of the planet being wiped out, leaving just the core of it. This was the first documented report of this kind of phenomenon. Planets such as this which orbit closely to their star can't endure these short distances and essentially melt. This one, just 4 million miles away distant from the star was warm and "puffed up"[32in its evaporating hydrogen atmosphere, leaving the trail of a comet within a narrow, 3.5-day orbit. The location of the planet was too close to Hubble to take a direct image of it however it could use the spectrum of imaging utilized in the event that the planet partially blocked light from the star and the emission of hydrogen dropped dramatically during those time points.

In the month of March, 2003 Hubble observed a "light echo"[33within the V8381 star, Monocertis. It suddenly increased in brightness to six million times the brightness of our sun. Then, the bright flare reflected off those dust cloud clouds. It briefly became the brightest stars in the Milky Way, but faded to near-obscurity after. Hubble recorded the most extreme light echo that has ever been seen, as if flashbulbs were reflected that are reflected in fog and a clear image of a CAT scan that resembles the three-dimensional structure of dust particles that surround the star was recorded. This image was an account of the star's unique explosion that was 20,000 light years away from Earth. Despite this distance it was able to release enough energy to brightly reflect light on the dust surrounding it before reflecting towards Earth through an indirect pathway that made it delay the time it took to reach us in the present.

Another episode of trouble not so comfortable as the mirror incident occurred in 2004. The most serious issues occurring with the Hubble during its early period was that of the Space Telescope Imaging Spectrograph. The device that separated light beams into their individual colors broke down four years into the mission, and did again, a few years afterward. A malfunction in the power supply necessitated an upgrade of the low voltage circuit board that was used to supply the power. In order to accomplish this, specially-designed tools were created.

A few researchers have ever had an chance to observe a collision between a comet at close range, however in July 2004 it was the NASA Deep Impact spacecraft Deep Impact intentionally created such an event when it fired an 820-pound projectile into the direction of comet 9P/Temple 1. Hubble's observations, including prior and post-exposures clearly show how Temple 1 appeared four times brighter than it was before, and with an outer cloud of gas and dust that surrounded the object, growing to 200 kilometers in dimensions. After about 62 minutes of dust and gas dispersal the debris field began to expand outwards in a fan-shaped shape and traveled at speeds of 1,800 km/h.

The wear and tear of the gyroscopes could lead to a mechanical breakdown and, as a result, Hubble operations were switched to a two-gyro setup to ensure the longevity of a third in August of 2005 and the life-span of observation for Hubble was extended by eight months. Hubble telescope extended until the year 2008, which was eight more months. The choice of two gyros rather than three gyros - which are the basis of the pointing system - was unaffected with regard to performance. The Hubble telescope requires to know where it is when it makes every observation, in preparation to find its next target. The three gyros used to supply those data, but that task was handed over to Fixed Head Trackers.

When looking up at Pluto on October 31st in 2005, two tiny moons that orbited the planet's dwarf were spotted through the telescope. They both wobbled in a erratic manner and, if one was capable of standing on the surface, it could not be able to determine the timing of the sunrise. The "cosmic dance that had chaotic rhythm"[34was most likely resulted from Pluto and Charon moving around one another in a constantly changing gravitational field. The shape of the moon's football added to the effects of "tumbl[ingto tumbl[ing] erratically."[35 The colors were awe-inspiring in that Kerberos was dark as charcoal briquettes, and the other moons were light like sand.

Then, two weeks later, the moons as well as rings of dust that resembled Saturn around the solar system Uranus have been discovered. The two moons were named Mab and Cupid The two moons have seen their orbits change significantly over the previous decades. Two dust rings that were not seen before showed bright clouds as well as an extremely high-altitude haze over the pole of the south. The images were captured by WFPC 2 the first occasion that such details were observed since the Voyager spacecraft's flyby.

The Hubble telescope has observed the impact of a comet on the Sun between April

18-20 in 2006. Comet 73/P/Schwachmann-Wachmann 3 broke into fragments as it approached the Sun. When it was at its closest the fragments were renamed alphabetically. Hubble captured pieces B and G just after massive bursts of activity. These fragments were pulled down through tail outgassing in the same manner as astronauts move using jetpacks. The comet's destruction was an orderly "hierarchical destruction"[36where smaller chunks of the comet accelerated away from the nucleus of its parent more quickly. Cometary nuclei are frozen relics of the solar system's early days.

While scientists have yet to come up with a definitive definition of what is dark matter's nature, Hubble did provide further evidence that it exists on the 31st. In cluster 0657-56which is also called the Bullet Cluster The Hubble Space Telescope witnessed a collision between two huge galaxies in a cluster. Photographs of hot gases in X-rays reveal the two red "clumps," containing most of the "normal" matter (baryonic). But, when hot gases travel across one another, nearly everything else appears "blue." Dark matter isn't slowing down because of the collision, and it does not interact with gas, itself or normal matter as long as gravity is not in the mix.

Comets' unpredictable behavior continued to astonish observers who depended on their telescopes. One "mystery comet" was seen on the 29th of October of 2007 when it suddenly grew by a million times over a 24 hour period. It had a tail of 400,000 miles, and was spotted close to the Trojan asteroids - remnants from an earlier system - and around Jupiter. This was the very first instance that that a new comet had been discovered in the area. The comet that was rogue came from the family known as Centaurs Ice bodies, which were found in the region between Jupiter as well as Neptune. The eventual outcome could be being drawn to Jupiter as well as the Sun. If it is ejected from the solar area, it may carry into interstellar space as a comet.

In the last quarter of 2007, a cloudy extrasolar atmospheric layer was observed in the orbit of an ethereal star. It was HD 189733b was a carbon dioxide-rich planet, which was the first time it was discovered within the atmospheric layer of a different star system. A crucial chemical signaling for extraterrestrial life this planet's temperature was far too high to be any form that could be recognized. Early observations showed methane and water vapor within the atmosphere. Carbon dioxide was of particular fascination to those looking for connections with biological life.

Additional "firsts" were to come the 8th of March, 2008 as an organic chemical was discovered in the methane atmosphere of an exoplanet this was the first instance of it being outside of our solar system. Utilizing spectroscopy to reveal the fingerprints of different chemical compounds was an "crucial stepping-stone"[37in assessing the prebiotic molecules that could be found in environments where life might exist. Further investigations confirmed "there is water."[38The location is located 63 million light years away from the Vulpecula constellation ("the little fox"). This planet is one type of "hot Jupiter" type, which is so close to the star's parent that orbiting can take one to two days. Its temperature is around 900 degrees Centigrade.

The Hubble telescope surpassed 100,000 orbits around the Earth on the 11th of August 2008. The telescope was two decades in orbit, travelling at a speed of five miles per second, and having completed 2.72 billion miles which is equivalent to 5,700 trips round trip towards the Moon. The day's mission included turning the camera toward it's Tarantula Nebula near Star Cluster NGC 2074, which is 170,000 light years away from Earth.

A Hubble telescope image of a star-cluster in the Tarantula Nebula

The visible light images of an exoplanet that orbits the Fomalhaut star were published in November that year. The Fomalhaut star system is located 25 light years away from Earth and the planet that was targeted that is visible light source close to the constellation PiscisAustralio was identified as Fomalhaut B. It was an appropriate location as an exoplanet could be a good candidate for analysis. The problem was that Fomalhaut B was able to throw Hubble an angle, before disappearing and disappearing in later exposures. Alarms were raised when the object failed to show the behaviour of a genuine planet. It could have been an unidentified cloud that was disguised as the planet, or an collision could have eliminated it. The orbit appeared odd even to the point of being eccentric and did not provide an infrared signal.

The most recent mission took place over a 12-day period that began in the month of May. After numerous delays, the spacecraft Atlantis carried a seven-member crew for the HTS to set up the Cosmic Origins Spectrograph (COS) and WFPC3 and a brand newly-built science computing system. An orientation sensor has been upgraded as was the insulation for three electronic bays as well as a gadget was added at the bottom of the telescope in order to allow de-orbiting in the event of closing. Alongside minor repairs were the substitution of six old batteries.

Hubble's lens was returned to the asteroid belt in February 2010, and pointed at the asteroid Vesta which is the second-largest of objects located between Mars in size and Jupiter. A chunk of Vesta fell into a huge fireball in Australia 30 years prior to Hubble's launch. In terms of size, Vesta, visible to the naked eye, is just surpassed by Ceres which is classified as dwarf planet. It was discovered near Earth in the year 2004 when Hubble has mapped its topographic surface and discovered that it stretches 530 kilometers across and has a layer of lava that has been cooled, covering an erupting mantle. The core is composed of nickel and iron and the lava was once flowing several kilometers beneath. A remarkable volume of hydrogen was found as well as extremely reflective patches of the beginning of its existence, the time it was flooded with liquid water, can be observed.

In 2011, Hubble's interest was redirected to dark matter, which is now identified as the force that is taking our universe to pieces. Large galaxy clusters acting like magnifying lenses within the space around them, Hubble constructed a sharp map of dark matter throughout the universe. This is the Abell 1689 galaxy cluster lay 2.2 billion miles away, and contained thousands of galaxies as well as trillions of stars. Hubble employed 135 lensed images of 42 galaxies that were background to determine the location and amount of dark matter. Then, Hubble superimposed the map of infrared dark areas, which were colored blue. The galaxy was blessed with plenty of dark matter at the time of its beginning and has carried some of it since.

The Hubble agenda was expanded to include a number of major projects on this Hubble agenda, such as Hubble's Cluster Lensing and Supernova Survey. In a period of three years the Hubble team identified 25 clusters that had high emission from X-rays were planned to be studied. The X-ray emission indicates a large amount of hot gas, which indicates these clusters could be huge. Hubble was able to determine the distribution of dark matter, and find evidence for the early formation of clusters.

In July of 2011, Hubble was well into 1,000,000 observations of science. That day, the HTS was looking for liquid water inside the atmospheric layers of planets located 1,000 kilometers away. The millionth photo was taken of HAT-P-7b, which is a gas giant that was larger that Jupiter that orbits a sun much hotter that the sun. So dependent was the astronomy planet on HTS that, in the end of December the tenth Hubble science paper was released.

One of the more significant subjects was an examination of Andromeda Galaxy on May 31 2012. Andromeda is currently on an collision course to the Milky Way, and the coming merger could be profound. It is believed that the Milky Way contains billions of stars spread across 100,000 lightyears of space. In the same way, Andromeda is much larger and twice the size of ours, with trillion stars locally, the largest and bright galaxy. The stars of The central region are abundant of heavy elements. In contrast to our old galaxies, which are generally older over 13 billion yearsold, Andromeda's stellar streams range from that are between 6 and 8 billion years old meaning that a cosmic act of cannibalism is taking place. It's vast and away, yet Hubble has taken a picture of the length of 61,000 light years and put it together into a mosaic of 7,398 images of 411 individual points. The image composite is part that is part of the Panchromatic Hubble Andromeda Treasury (PHAT) Program.

Hubble was back on Pluto on the 11th of July 2012, and found five moons orbiting this dwarf planet. The new orbiter was only six to 15 miles in size, with an irregularly-shaped diameter. The track covers 29,000 miles within a sequence of "nested" orbits. Fourth moons were first discovered in the year before when NASA's New Horizon spacecraft prepared for an orbital flyby.

The most recent composite image to depict all cosmic events of one sky patch was announced on September 25 on 2012 as"the Hubble Extreme Deep Field (XDF). The first glimpse revealed the way that the early galaxies appeared to look like within the visible spectrum. There were also some remain visible within the "dark age." This was the time between the initial release of the cosmic background, and also the point when the development of structure of the universe caused the collapse of gravitational gravity of objects, leading to the formation of the stars. Hubble captures the universe in only a small fraction of its present age. Advanced Camera Survey and WFPC3 made the images in an ensemble effort lasting more than 10 years.

The Hubble telescope is celebrating its 23rd anniversary within space, the Hubble telescope was able to study in detail the Horsehead Nebula on April 13 2013, which is a tiny portion of a complex of star-forming stars located in the Orion constellation. The constellation has the possibility of disintegration within 5,000,000 years. It is believed that the "Horsehead" is hard to spot using a telescope that is personal, however the Hubble images in infrared are very popular with the general public. A dark cloud of molecular matter, located 1500 light years away is classified informally to be Barnard 33.

The Horsehead Nebula

The last months of 2013 saw an asteroid study in November, where the main body was accompanied by six comet-like tails that resembled comets. But, December was more significant because the plumes of water-vapor were found within the Jupiter's Europa in the cold south the polar region. Prior studies suggested the possibility of an ocean under Europa's crust. If the link to subsurface activity is established scientists can now directly examine possible habitats for life without drilling into the ice. In addition, as was observed through the Cassini mission The Hubble telescope was stretched into its limits to observe an emission of this magnitude. The moon's fractured, long cracks, known as lineae could also signify water vents. However, at the most close places to Jupiter the moon does not vent. This could indicate that there is a tidal flex beneath when the plumes drop back towards their surface, at an elevation of about 125 miles.

The latest installment of the history of composite images was announced on January 7th the 7th of January 2014. It was a galaxies cluster Abell 2744 as part of an ongoing program that surveyed the super-deep field by taking long exposures. These included the most faint images ever captured. The image was dubbed"the First Frontier Field Image, it was displayed during the annual conference of the American Astronomical Society in Washington, D.C. Abell 2744 contains hundreds of galaxies when they were discovered in the beginning of the universe. The massive gravity of background clusters was used to create "gravitational lensing,"[39that stretches space, magnifying and brightening distant galaxies that were once dim.

The composite also serves as an experiment that tests whether Hubble's incredible quality as well as Einstein's Theory of Relativity to reach the first galaxies. Some of the backgrounds are between ten and 20 times bigger than how they normally appear. In the absence of gravitational lensesing the image will have to be "smeared," stretched, and then duplicated across the field. Hubble can observe dwarf galaxies that are 01.100th that of mass within the Milky Way. Plans are being made to revisit them however, the optical wavelengths in the visible as well as infrared will switch to the second view.

An asteroid that was disintegrating was discovered at the Hawaii's Keck Observatory and turned over to Hubble in March of 2014. The asteroid was photographed when its fragments dragged apart at a speed of one mile per hour, removing collision as the cause. The ten fragments are now falling apart. The majority of dust and ice comets break apart close to the Sun however, one of them had not been observed to do so in the Asteroid Belt. The largest pieces were about approximately four football fields long. The cause of the disintegration due to the fact that the Sun's strongest heat was not in the vicinity the area, could have been due to sunlight which caused an increase in rotation . This could have led to a weakening inside, as is also seen in other instances--the fragments succumb to the force of centrifugal force, and then break off.

The changing environment of Jupiter's planet required the inclusion of Hubble's agenda during May of 2014. "The "Great Red Spot" seemed to shrink and change in size from circular to oval. In the past, the dot was able to contain three Earths within its radius however, it now was just one-third of the planet's smaller size that was only 10,250 miles wide. The shift in shape is due to tiny current eddies that are joining the storm.

New targets for exploration were planned to be discovered in 2014 within the Kuiper Belt as a boost for the New Horizons spacecraft, preparing to depart from Pluto to follow. Hubble has captured images of the surface of the planet, where ground instruments couldn't. Following it was discovered by New Horizons, Hubble found an additional target for a flyby and after scouring the depths of the sky the tiny, icy world was found within the same region that were beyond Neptune's orbit. The combination of Hubble's Hubble Telescope and space probe that was traveling at 32,800 miles an hour confirmed the previously unexplored Ultima Thule.

Finding Pluto's moons Pluto prior to their arrival on New Horizons was timely. New

Horizons would have discovered the moons, too however, only two months before the launch, with no time to preparation. The arrival of Ultima Thule was predicted for New Year's Day in 2015.

In mid-January, a panorama image of the Andromeda Galaxy was published the mosaic comprised of 7,389 images, including 100,000,000 stars and thousands clusters. It was the biggest ever of this kind of image.

The first supernova predicted was discovered on December 6 of 2015. It was a supernova that exploded within a cluster, which caused it to be visible in various places at various time. Timelapses of the spiral galaxies, NGC 2525. They show an aftermath of the explosion, with the white dwarf "slurping"[40in the form of material from a companion star , and the explosion exploding in a thermonuclear blast. The white dwarf is located around 70,000,000 light years away The brightness and speed of the fade allowed for the measurement of distance.

It was reported that the "Cosmic Distance Record" was broken once more 3 March 2016 when Hubble discovered a galaxy just 400 million years from the Big Bang. The bright

GN-z11 galaxy, in its current form, as seen 13.4 billion years earlier was the most distant that was found, beating earlier records by 13.2 billion. Given the intensity of the new galaxy, it suggests that "other galaxy clusters that are unusually bright in earlier images are in awe-inspiring distances."[41[41]

The moon of a new phase was observed in orbit around the tiny world Makemake located in the Kuiper Belt on April 26 2016. The planet was named after the god of creation, people of the Rapa Nui inhabitants who live on Easter Island. Makemake is the second-highest-glaring dwarf planet that is icy and its moon measures only 100 miles in size 870 miles long and incredibly small. Its predecessor, the WFPC 3 employed the same procedure used to create the moons of Pluto.

When investigating the atmosphere of exoplanets back in the month of February, seven Earth-like planets that orbited Trappist-1 were found. A few of them resided in an area that was habitable according to the spectrographic study. One is not surrounded by cloud of hydrogen that is puffy. Hydrogen is the greenhouse gas that covers planets orbiting too close the star's parent, and the other. The results favor cloudless orbs in environments that are more closely aligned with Earth, Venus, and Mars.

At the same time as universal expansion and the elements found on exoplanets, evidence of gravity waves that extend 130 light years away from Earth. They cause ripples in space caused by violentand energetic events occurring in the universe. Hubble identified the origin of the resultant "kilonova,"[42which is a stellar explosion that results from the collision of two objects that are compact. Neutron stars may produce gold, plutonium along with other components, theoretically creating gravitational waves as well. The first gravitational waves that were discovered during October of 2017 followed by short gamma blasts from NCG 4993 that travelled a long distance. The pulse of a massive blast of light was detected all over the globe. This was the first occasion that we have seen both gravitational and light waves arising from the same source. The distantst star ever discovered was discovered in April 2018and known as Icarus. Its light source was magnified by the interplay of galaxies, in the now-common method of gravitational lensesing. The formal name for the star is MACSJ1149+2223. Lensed Star 1. By gravitation lensing, Icarus is the very first instance of being able to see an "magnified individual star,"[43 According to Patrick Kelly of the University of Minnesota.

After having observed interstellar objects as well as phenomena that were observed in the night sky, we finally got to see our solar system on June 27 the 27th of June, 2018. The first interstellar object to be recorded Oamuamua was spotted in the solar system before experiencing a sudden increase in speed. In this manner it behaved like a comet, dispersing gaseous matter. The discovery was made by the huge telescope in Hawaii Its tracking was financed through the Near Earth Object Observations (NEOO) the organization that monitors new threats. On the 17th of September the 17th, it "slingshotted"[44over it's Sun with a speed of 196,000 miles an hour. The object's origin is probably that of the constellation Vega. Its brightness varies by 10 times the amount when it turns around its direction. There is no known comet or asteroid in the solar system has ever been so diverse.

On the 3rd of October 2018, 2018, a potential exomoon was detected outside of the solar system, larger than Neptune and with the planet that is more massive than Jupiter. The sun-like planet orbits a star that is 7,800 light years away from Earth. The moon in question is ten times larger than all moons and planets in our solar system. The research team included an "extraordinary claim,"[45 and the cautionary tale was issued. The information was given to Hubble and, forty days later the evidence was confirmed. The crucial next orbit across the the earth started about 78 minutes earlier than what was predicted which suggests that another planet is "tugging" towards it. The moon and the planet that orbit each other could be the reason. David Kipping of Columbia University said that the research team of his was "unable to discover a theory that could explain all the information we have."[46[46.]

The final chapter of Hubble compositions came out in the month of May of 2019 and was named"the Deep Field Image Legacy. The area of sky that was examined was almost the size of the full moon. The image is comprised of 7,500 exposures that were taken over 16 years. It includes 265,000 galaxies as well as 13.3 billion years cosmic history.

Hubble continues to expand beyond its anticipated lifespan far in the 20th century. On September 13, 2019 water vapor was found on an exoplanet that is in an area known as the habitable zone. The planet orbits around a tiny red dwarf, which is located 110 light years away, in the constellation Leo. It could be too extreme for us, but hundreds of possible "super-Earths"[47are being studied.

Galaxies are among the most effective laboratories for studying the long-term world as well as Hubble himself has declared galaxies to be "markers of space."[48Because of a deficiency of adequate optics and computing power that is weak, as well as the lack of any delivery system Hubble telescope is the best choice. Hubble telescope gives a glimpse of the beginning of space, time and the start of life. At one time, it was thought impossible but scientists continue to receive photos from the very beginning of space and time and have the ability to capture the photons.

The Hubble Space Telescope began to fail in the year 1990, no one would have thought that it would last longer than it. Space Shuttle Program. Space shuttles' versatility allowed it to carry out various types of missions while in orbit, ranging from servicing spacecrafts like the International Space Station to conducting scientific experiments for weeks in orbit. Hubble telescope Hubble telescope was constructed using the space shuttle's capabilities in mind, so it could be used in future shuttle flights to fix and maintain the telescope.

But budgetary concerns as well as a shaky economic situation made NASA and space exploration targets of possible cuts in the decade of 2010, resulting in doubts about America's space policy in general. NASA predicted that 2009's servicing of its telescope was its final one, and given the space shuttles' retirement and no spacecrafts scheduled to replace them which means that once they're retired, servicing missions for the HST are not in the possibility.

In the absence of future servicing missions there is no matter of if the Hubble telescope's components will fail, however, it is when. The telescope was scheduled for a maintenance mission every three years. The instruments and equipment failing more frequently. And while the telescope was fitted with the latest instruments, orbiting Earth's orbit around Earth in space is extremely damaging to the instruments. The instruments have to work with each other in order for the telescope to operate to its fullest potential. Additionally the telescope is designed to be de-orbiting the Earth and disappear from the sky in just 20 years.

Technology advancements have enabled telescopes and observatories that are located on the surface to expand their scope, but NASA plans to launch two telescopes into space. The most popular telescope will be that of the James Webb Space Telescope named for NASA's director in its Apollo missions. The telescope will feature more powerful quality optics than Hubble telescope, however the main benefit will be its ability to detect infrared radiation. It is important to note that the Hubble telescope was not designed to take advantage of infrared technology. However Wide Field Camera 3 was capable of detecting infrared radiation. Infrared helps telescopes see more stars in the deep space. With improved infrared capabilities telescopes like the James Webb telescope is expected to be able to get more detailed pictures of space than the Hubble telescope.

There is a reason why the James Webb telescope or other telescopes could eventually become superior and more efficient over the Hubble telescope however, the Hubble telescope laid the foundation for the future exploration of deep space and set the standard for operating the space telescope. Newton once wrote "If I've seen more than that, it was by standing upon the backs of gigantics." Following the publication of more than 15,000 scientific papers which are based on Hubble Space Telescope's research and the collection of stunning images Future missions will surely depend on the telescope's years of data collection as the primary source.

## Chapter 4: The New Universe

This image captured from The Hubble Space Telescope depicts dark pillar-like structures often referred to as "The Pillars of Creation," which are actually super cool hydrogen gas interstellar as well as dust that also act as incubators for new stars within the Eagle Nebula. This color photo is crafted from three distinct images taken with different light sources. kinds of atoms. Red is the color of sulfur atoms, while green is of hydrogen and blue is from oxygen, on April 1st 1995. (AP Photo/NASA)

By using the power of photographs The Hubble Space Telescope uncovered the mysteries in the world.

For the past 19 years, Hubble has shown the massive violence of galaxies that crash as well as observed star births and deaths, shared cosmic lessons, and created comic relief.

In Hubble's photographs believers see the hand of God Non-believers can see astronomy at work, and artists find galaxies worthy of galleries.

In the near future, Hubble is set to receive its fifth and final overhaul. If everything goes according to plan then spaceship Atlantis will take off on Monday (January 28 2013, 2013) to take a trip to the telescope's orbit, which is 350 miles high above Earth. In five very long spacewalks astronauts will repair and replace damaged instruments, and also install a new long-gazing camera and the last thing they will do is say goodbye to Hubble. If all goes well, Hubble will get another five to seven years of its life and will be remote-controlled and sucked into a muddy grave.

Hubble does more than just tell how the Universe works. It also has its own tale which includes failures and redemption.

Hubble's senior Hubble researcher Mario Livio rhapsodized about the tragedy of Hubble's own tale, "turning something that could be the most scientific disaster into the most significant scientific triumph."

Following its launch into space in the year 1990 The Hubble Space Telescope was stuck with blurred vision due to the fact that its mirror was not quite right. It was the basis of jokes in late-night comics. An editorial cartoon claimed its creator was named Mr.

Magoo, a nearsighted cartoon character. It appeared to be a massively over-budget error.

After it was repaired three and a half years later using a new set of glasses Hubble lost its skewed image. Hubble began producing images with far-sightedness of space, which appeared to be more like art than the science of astronomy.

Hubble was able to determine the universe's age at 13.7 billion years. It also explains the nature of the universe and also show where it's headed. Hubble's images indicated that, as a planet Earth might not be the only one. One image of galaxies with warped edges gave a visual proof of Einstein's theories of relativity generalization.

"Hubble actually allows our human brains and spirit to journey light years, even billions of light years," said NASA science head Ed Weiler. The photograph "Hubble Ultra Deep Field" depicts a time when the universe was 700 million years old. therefore the stars that make up it are 13 billion light years away. One light-year equals 5.9 trillion miles.

A new camera that will be fitted in this flight will allow astronomers to see at an additional

200 million light years further back, according to Hubble director of science David Leckrone. He added that if all goes as planned for repairs, Hubble will be at its most precise level ever.

This image is a representation of what scientists from the space agency call"the Hubble Ultra Deep Field. In this image, Hubble Space Telescope Hubble Space Telescope looks the most distant we've ever seen into the universe, taking the light spectrum from the 13 billionth year ago, when the universe was just 700 million years old. 2004. (AP Photo/NASA, File)

It was an image from the Hubble image taken in 1995 which has forever restored the telescope's initial reputation. The image featured the Eagle Nebula. It was spectacular with gorgeous colors and stunning clouds where stars were formed. NASA described it as "the world's pillars."

The public, who used to snicker at Hubble and chuckled at Hubble, was now smitten.

Hubble has captured 570,000 images and, while some of them capture the birth of planets and stars and other objects, others document the opposite final stage of life which is violence and death on a grand scale.

"We have 20 beautiful pictures of stars that look similar to our sun's in its dying phase," said Hubble astronomer Frank Summers. "They are absolutely stunning. It's awe-inspiring to imagine that stars that look identical can end up in so many various ways."

The coil-shaped Helix Nebula, also know as the "Eye of God Nebula" is among the most massive and detailed images of the night sky ever produced. The composite image is a seamless mix of sharp images taken by NASA's Hubble Space Telescope together with the vast image of Mosaic Camera that is located on the National Science Foundation's 0.9-meter telescope located at Kitt Peak National Observatory near Tucson, Arizona. The image displays a beautiful grid of filamentary "bicycle-spoke" elements that are embedded within the vibrant red and blue gas ring. Nearly 650 light years away The Helix is among the closest planetary nebulae Earth. A planetary nebula can be described as the glowing gas that surrounds the dying Sun-like star, on May 9 2003. (AP Photo/NASA)

As the ages began to catch up with Hubble -- which was intended to last for between 10 and 15 yearsit was NASA initially decided that the telescope was going to gradually die. A repair mission for astronauts was considered to be too risky in the period following in 2003's Columbia shuttle crash that killed seven astronauts. But eventually, public opinion and political leaders convinced NASA to change its position. The fervor and promise of stunning images trump estimates of risk and cost.

"It has become an iconic part of American daily life," said Weiler, Hubble's public face. Hubble since its inception.

While most people love Hubble from distance, those who have seen the Hubble up close discover it to be a persona.

"It's nearly impossible not to feel like Hubble is a living thing," said astronaut John Grunsfeld who has worked on fixing the telescope twice before and is set to climb inside Hubble's hood for the third time. "It's an ordinary satellite however, once you've gotten involved with the program and become completely enthralled It's very simple to begin adding personalization to Hubble.

"I am feeling like ... that I'm heading to see an old friend who I haven't seen for many years that is going to be a more weathered, and a older," Grunsfeld said in an interview in the fall of. NASA hasn't been to Hubble for seven years , and expects to see indications of wear and tear which could include holes from space debris.

The telescope isn't inexpensive. NASA believed that it could build Hubble at a cost of $300 million. However, it costs over five times the amount. With all the upgrades and fixes and years of use the total cost is expected to be in the vicinity of $10 billion when it expires, and nobody is complaining about that cost, Weiler said.

Astronomer Livio stated that certain photos recall abstract paintings. The colors -which are incorporated to the ground, since the cameras can only shoot black and whitethey can appear ugly. But then again, so is the universe.

"This is artwork on a grand scale" Astronomer Summers stated.

**Chapter 5: 1990 Introduction**

The Space Shuttle orbiter Discovery lifts off from Launch Pad 398 in Kennedy Space Center in the morning. It is with a crew of five along with Hubble Space Telescope. Hubble Space Telescope on April 24 1990. (AP Photo/Paul Kizzle)

Seven years after a gruelling delays After seven years of frustrating delays, the Hubble Space Telescope is now placed on board the spaceship Discovery and is ready to set off on its journey to search the space's deepest depths.

The $1.5 billion observatory is an astronomical leap in astronomical magnitudes the largest ever since the first time that we used telescopes made by Galileo from 1609.

"We're aware of a lot We think. Maybe we don't know everything. Maybe we'll find out" said astronaut-astronomer Steven A. Hawley, one of five members of the team which is scheduled to begin the launch of Discovery the next day, Tuesday (April 10 in 1990).).

One of the mysteries Hubble could aid in understanding are pulsars, quasars as well as black holes along with the precise source of the universe, as well as the possibility of the presence of solar system with other stars as well as life forms.

However, the telescope's most astonishing discovery during its 15-year journey could be one that isn't yet thought of, according to Eric Chaisson, senior scientist at the Space Telescope Science Institute at Johns Hopkins University.

The history of science has proven that the most exciting discoveries are ones that the human imagination has never even thought of in the first place,'' he stated. "We will demand that the Hubble explore areas of darkness to explore areas that no one has ever before looked and hoping to uncover the unimaginable."

From its perch of 380 miles up from the ground, from its vantage point 380 miles up, the Hubble will be devoid of atmospheric distortion , and thus will be able to discern light that is of all wavelengths, including ultraviolet. The telescope is capable of detecting objects that are fifty times more faint and with 10 times better clarity than the most renowned observatories on Earth.

Astronomers are hoping to study galaxies, stars, and stars in such a distant location that the light they emit has been travelling towards Earth since 14 billions of years.

According to some theories, the universe was formed 15 billion or more years ago through a massive explosion dubbed"the Big Bang. The theory is based upon the findings of the late American Astronomer Edwin P. Hubble that the universe continues to expand.

The mission of the shuttle that will send the observatory named after that pioneer in orbit has been anticipated by NASA and scientists across the globe for over 10 years.

Hubble Space Telescope Hubble Space Telescope originally was set to the launch date in 1983 just five years after receiving funding from the government. Problems with the launch system delayed it for several years. The launch was delayed until 1989 due to The Challenger explosion on January. 28th, 1986.

Concerns over the solid rocket's boosters delayed the mission again, this time to mid-April, so the booster's lower right section could be repaired.

The minor issues continued to pop up, including the discovery of a few dozen midges, small insects that resemble mosquitoes, had entered the payload prep area on the pad for launch. Engineers were worried that the bugs could cause damage to the 23,250-pound, 43-foot telescope, and the installation was delayed while the midges were taken care of.

Then, a pipe that was damaged caused water to splash onto electronic equipment, and shut off air cooling. Certain computers that are sensitive to heat were shut down temporarily stopping testing of the telescope's science instruments.

"Every new issue causes you to be a bit tight. We've gotten through it" stated Bill Taylor, the telescope's chief engineer.

When the telescope was first introduced on the Kennedy Space Center last October The engineers had made use of the delays numerous times to modernize the telescope and allow it to be simpler to take care of in space.

The telescope, valued at $1.5 billion is the highest-priced spacecraft that has never been manned by NASA. Additional $500 million has been earmarked to store spare components, train astronauts, and develop computer programs for the operation of the telescope. It is the European Space Agency is paying 15 percent of the cost.

Discovery's five astronauts have said they'll be unable to breathe till the Hubble is secure in space. The tense waiting will take longer.

"Their hearts are going be pounding, they're going be fluttering with butterflies, and they're probably not going be happy for a few weeks after the release and they begin receiving information in the mail,"'' said Marine Colonel. Charles F. Bolden Jr. The pilot of Discovery.

"Hawley will utilize the spacecraft's robot arm that measures 50 feet long to take this Hubble telescope from Discovery in the second of five days. NASA would like to deploy the telescope as early as it is possible during the flight in order to limit the possibility of contamination of the 8-foot primary optical mirror.

We're taking every precaution to to protect any possible issue that might arise while deploying the telescope" stated William

Reeves, NASA's lead flight director for the mission.

The mission specialists Bruce McCandless II and Kathryn Sullivan experienced spacewalkers will assist to take off, fix or remove the telescope, in the event of need. The pressure inside the cabin will be reduced 3 hours prior to the flight , and then remain lower for a few days to cut down the four-hour reaction time that is normally needed to complete the spacewalk.

Space Shuttle Discovery Space Shuttle Discovery lifts off from launch pad 39b and carries the Hubble Telescope located at the Kennedy Space Center in Florida, USA, April 24 in 1990. (AP Photo/Steve Helber)

Discovery Pilot Charles Bolden of the Hubble Space telescope flight on March 22 1990. (AP Photo/Ed Kolenovsky)

It is estimated that Hubble will cost around $200 million annually to maintain and operate on the orbit.

In about a couple of months, the information for the tests will be transmitted through satellites to Space Telescope Operations Control Center located at NASA's Goddard

Space Flight Center in Greenbelt, Md., as well as the Space Telescope Science Institute in Baltimore.

Six scientific committees selected 160 submissions from astronomers around the world which included five amateurs for the instrument's inaugural year of observations that started in December. Ten times more observation time was demanded than the institute was able provide. Each year, new requests will be considered.

A spacecraft will be scheduled to the telescope to repair it and, as its orbit will inevitably degrade in time, the telescope will be boosted to an appropriate height. Additional maintenance missions are scheduled every three years. Emergency work will have to be incorporated in the already busy flight schedule, said Discovery's commanding officer, Air Force Col. Loren J. Shriver.

More than likely, you'd need to commit an entire mission due to the altitudes are high in the mountains, Shriver said. Shriver stated.

Discovery's ascent of 38 miles would be the longest that a spacecraft has ever climbed. The astronauts will take the telescope back if they is not able to be safely deployed. However, that's an option last resort, Hawley said.

"We'll do everything we can to keep the space telescope intact, so that it is able to be used to be used in a mission that is successful, Hawley said. Hawley declared.

*Astronaut Steven A. Hawley, July 24 1980. (AP Photo/NASA)*

Questions about cosmic physics that have puzzled humans since the first time they looked at the night sky could be answered once Hubble Space Telescope is launched. Hubble Space Telescope is launched into orbit, forming the most keen celestial eye.

## Chapter 6: Keenest Eye Space

Humanity has never had the an opportunity to view the universe as clearly as we do now,"" declared NASA director of science Lennard Fisk. It is sure to be awe-inspiring to us."

Hubble which is scheduled to be launched on Tuesday (April 10 1990) by the spaceship Discovery will offer an overview of the universe that isn't filtered or obscured by Earth's atmosphere.

After orbiting the telescope of 12 tons will reveal sharp the dazzling vistas of stars that are currently blurred or not visible from Earth. Black holes. Infant galaxies. Planets which orbit distant suns. Stars in the birth, and at death. The daily details of Earth's solar system's neighbours. There are clues as to the origins of the universe as well as the development of millions upon billions on billions of stars. These are all subject to study through the humankind's brand new keyhole to the stars.

"This is a place that (people) could expect fundamental findings that will change the way they think," said Fisk. The scientists could start to understand the origins of our universe. Hubble may be an important turning point for humanity's perception of its self and its place within the universe."

Astronomers, primarily will be expecting the $1.5 billion telescope to uncover surprising discoveries. A lot of scientists who've built careers of speculating on the things they cannot see will be putting their reputations at stake when the telescope provides new insights into the universe.

Theories could be discredited however, that's how science operates.

"We'll be adding little pieces to our theories, and possibly some major pieces according to James Westphall of the California Institute of Technology. But more likely we'll throw a lot of pieces in the trash bin.

"One of our greatest pleasures," said he "will be doing one or two theories."

Some of the issues to which scientists are seeking answers include:

When did the universe first begin?

Hubble is a time machine that frees our curiosity and allows us to go back 10 billion years back to an era where there were remnants from The Big Bang, the massive explosion believed by astronomers to have started the universe.

It was at this point the hydrogen, Helium, and other elements of the basic kind emerged from the cooling debris, and then began to join to form galaxies, stars and stars.

The light that was created during this transitional time in our universe is scurrying throughout space since then but it's too dim to be observed through telescopes located on Earth.

Hubble will be able to capture the light, and astronomers will be able to interpret the message, searching for clues about fundamental phenomena that have intrigued humans ever since the beginning of time.

"This will provide you with an amazing glimpse into the very first seconds of the universe's existence"," stated Jack Brandt, a University of Colorado Astronomer.

How big is the universe?

The light emitted by Hubble's almost perfect mirror, transmitted to Earth and scrutinized by experts, may give the most accurate estimate of the size and form that the Universe. The telescope's measurements of distances to specific stars will give a measure that could later be utilized to determine more precisely distances to extremely faint stars as well as galaxies in the distance from us.

Because distance in space is equal to the length of time it takes, such measurements can assist scientists in arriving at more precise estimates of the time when it was that the Big Bang occurred. They will also be able to better identify the speed that the universe is expanding out of the place at the time of the Big Bang.

Edwin P. Hubble, the American astronomer to who the telescope was named, invented the formula used by scientists to calculate the speed at which galaxies and star clusters move from one another. Now , the telescope will improve the calculations of the person for who it was named.

Are there any people out there?

The majority of scientists believe that in order in order to sustain life beyond Earth it must exist on planets that are not part of our solar system. To date, no such planet has been discovered, but some astronomers believe there is some evidence to suggest that these planets are present.

Hubble is sharper than 10 times sharper than the other telescopes on Earth is likely to locate planets that orbit distant stars.

Most likely, scientists believe that the planet must be as large as Jupiter which is the largest planet of the Solar System before Hubble could even detect it. Even in that case, it will not be an actual sighting on the surface of the world, however rather a discovery based on an obstruction, or occultation of light from the outside of the planet. The existence of a planet could be detected by a gravitational phenomenon that is that is detected by the nearby star.

Discovering planets around other stars, according to Robert Brown of the Space Telescope Science Institute located at Johns Hopkins University in Baltimore is the closest thing that astronomers can do towards the golden fleece."

Dr. Edwin Hubble, Mt. Wilson Observatory astronomer, in this unpublished photo. (AP Photo)

The mere existence of planets orbiting other stars could be groundbreaking," said Brown. "It could alter our sense of home and raises questions about the diversity and diversity of our lives."

Even if a new planet is discovered but the telescope can't determine whether it is inhabited.

What's happening on planets that orbit Earth? Hubble's powerful optics can detect heads or tails even from miles away. The same visual power can allow researchers to investigate Jupiter, Saturn, Neptune and the rest of the solar system's neighbors through the sharpness captured by cameras aboard the Voyager spacecraft which journeyed to these planets.

Over the course of 15 years, Hubble will be able to take thousands of photos of certain characteristics of distant planets, permitting astronomers to monitor month-to-month developments. Hubble will be able for example, to analyze the evolution of the blue spot that is visible on Jupiter and to track the paths of objects within those rings on Saturn as well as to detect eruptions of volcanoes on Io the moon of Jupiter. moons.

What is the process by which galaxies, stars as well as black holes and clusters develop?

Just 20 of the close galaxies have studied in depth, leaving more than 100 billion galaxies being seen as fuzzy images of light.

Hubble is expected to examine hundreds of thousands of stars and will be able to view more clear the stars that have already been observed by Earth.

Scientists who have been forced to think about star formation, will be able observe the process in motion by using Hubble.

The telescope might also allow for the study of this strange celestial object called the Quasar. The mysterious, distant "quasi-stellar object" have been believed to be related to star formation , or perhaps they are radiation or light created by the powerful gravitational force that black holes generate.

Black holes are believed to be objects that have so much gravitational force that light, not even light, escapes. They are believed by some, to be at the heart of galaxies. Utilizing Hubble to study quasars as well as other black holes, astronomers could confirm their existence , and even determine their significance in the development of stars.

Another interesting object that is being investigated by Hubble is binary pairs. These are two stars , or two galaxies, which are formed in such close proximity that they swap material in a massive gravitational competition.

Hubbles Hobnob Cape Canaveral

The Hubbles are expected together alongside the Hubbells and possibly two Hubbels for an evening of family celebrations to celebrate the upcoming launch of Hubble Space Telescope. Hubble Space Telescope.

They're all related to the astronomer Edwin Powell Hubble, whose discovery in the 1920s of the universe was expanding was the basis for The Big Bang theory of creation.

Although he might not be the smartest or most wealthy Hubble, Hubbell or Hubbel who lived, he will certainly be one of the most well-known when the spaceship Discovery launches to launch Hubbell's $1.5 billion observatory named after him.

There's never been a family with this kind of reason to have an reunion",'' explained Harvey Hubbell IV, 60, a distant family member as well as a former business owner of North Palm Beach who is responsible for the reunion.

"What we were looking to do was to honor Edwin. Edwin has been acknowledged by NASA and this provides us with an opportunity to gather and study about Edwin and transfer the knowledge for the next generations", he added.

Around 200 relatives from Richard Hubball, a 17th century immigrants from England will begin arriving on the weekend. The event is described as a mini-reunion because the official gathering of the family occurs every two years, and this is a year of off.

Edwin Hubble, who died aged 53, was a astronomer who studied galaxies and stars in the 1920s at Mount Wilson Observatory in Pasadena, Calif. His survey revealed a vaster universe than previously thought. Hubble concluded that galaxies were dispersing from Earth with speeds equal to their distances which supports the idea that the universe began about 15 billion years ago as a result of the event that is known as The Big Bang.

There were additional relatives.

Harvey Hubbell's father, Harvey Hubbell, invented the electric socket that pulls chain and the duplex standard wall socket. The company he founded at the age of 1888

Connecticut has evolved to become Hubbell Inc., a Fortune 500 company.

It were Carl Owen Hubbell, the Hall of Fame pitcher who played for the New York Giants from 1928 until 1943. There was also John Lorenzo Hubbell, who created his first trading station within the Navajo reservation in the northeastern region of Arizona in 1876. The one located in Ganado, Ariz., is now run under the National Park Service.

The family reunions we have attended tend to be for retirees as well as the older generation. The interest is growing among younger generations according to Robert L. Hubbell, who will be taking two of his grandsons to this reunion in Falls Church, Va.

Robert Hubbell, a 71-year-old retired Foreign Service officer, is the president of the Hubbell Family Historical Society, that has more than 400 members.

The Dr. Edwin Hubble, space exploration astronomer, appears alongside his colleague Dr. Richard Chase Tolman, right famous mathematician and model maker of the proposed 200-inch telescope to be built for California in this unpublished photo. Hubble's observations and Tolman's calculations that helped Alfred Einstein change his mind regarding the Universe. Telescope mode was on display during the Summer meeting of the American Association for the Advancement of Science in Pasadena. (AP Photo)

Alongside all the Hubbells attending, a few family members from Edwin Hubble's maternal side are planning to join in.

Lena James Jump, 75 who is a third cousin on the mother's side will be in attendance with her three sisters.

It's going be amazing and wonderful," declared Mrs. Jump, a retired teacher of Marshfield, Mo., the birthplace of the astronomer. "I say if I die, I'll die happy."

National Aeronautics and Space Administration representatives will visit with the family on Monday. The next day the chartered buses will transport the family to a designated viewing area at the Kennedy Space Center, provided the launch time is within the timeframe.

A pool party is scheduled on Wednesday afternoon (April 10 in 1990) to be and will be followed by a visit to Space Center on Wednesday. Then there will be lunch at near Patrick Air Force Base on Thursday.

If the mission gets delayed beyond Thursday, we'll throw an event prior to the launch party to commemorate having everyone together and to give them Godspeed and success, Harvey Hubbell said. Harvey Hubbell said.

This will be the first time that most of the family will be able to witness an actual launch of a spacecraft. However, the closest living relatives of the astronomer and two sisters won't be attending.

Helen Hubble Lane, 91 as well as Betsy Hubble, 85, are ill who will watch telecast coverage from their home located in El Paso, Texas.

"We've always been interested. We're extremely excited to find out more," Betsy Hubble said.

The wife of the astronomer, Grace died in the year 1974. The couple did not have children. Two sisters and three brothers are deceased.

Betsy Hubble said Edwin went higher than any of the siblings. Edwin knew precisely what was important to him. This was his primary goal in his life."

Discovery's Astronauts Return for a Second Go

The five Discovery astronauts returned on Sunday (April 22 in 1990) for another attempt to launch the shuttle using NASA's most valuable and well-known payload that is Hubble Space Telescope, $1.5 billion Hubble Space Telescope.

Discovery was scheduled to take off around 8:31 a.m. EDT on Tuesday (April 24 and 25, 1990).).

We're sure that everything is going to be smooth this time around, said Discovery's commanding officer, Air Force Col. Loren J. Shriver.

"We'll be out well on Tuesday morning, and in the event that we don't then we'll try until we get it. It's the nature of the game in this situation", Shriver stated.

NASA testing director Mike Leinbach said Sunday that the countdown was going without a hitch and that the shuttle appeared in excellent state of repair. The countdown began on Saturday afternoon.

A malfunctioning power unit caused the first launch attempt be scrubbed just four minutes prior to launch on April 10. This unit has been replaced by an entirely new unit, and tests proved to be in good working order.

"I hope we can be able to get the shuttle off its pad by this point around, Leinbach said.

70% probability for favorable conditions was anticipated at the time of launch and low clouds the biggest concern, according to Air Force's Ed Priselac, shuttle weather officer. A cold front that was weak coming from the north was anticipated to cross the area on Monday night.

The outlook is significantly better on Wednesday as well as Thursday Priselac said.

The Discovery astronauts arrived at Johnson Space Center in Houston on Sunday afternoon, the spacecraft Columbia was being moved to the landing pad 1.6 miles away from Discovery.

This is the second time that both pad launchpads for shuttles on the Kennedy Space Center have been used at the same time. It was first January 1986. It was just 16 days following the time Columbia took off Challenger exploded.

Columbia commanding officer Vance Brand said he does not believe that the National Aeronautics and Space Administration is going too fast with his spacecraft. NASA anticipates that it will begin the launch of Columbia with an observatory for astronomy called Astro by May 16, a week later than originally planned due to Discovery's two week delay.

In certain situations it may be an issue" if there are the shuttles at both pads Brand said. In this particular instance the system is working."

NASA will have until the weekend to launch Discovery up and running. After this, all launches are to be stopped for eight days in order that Hubble's batteries can recharge.

Six nickel hydrogen batteries are expected to provide power to Hubble starting when Hubble is removed from the Discovery's electrical grid until the two energy collecting solar panels are installed the space station.

The Discovery payload compartment was closed late on Saturday night, dramatically decreasing the chance that the mirror could be contaminated. polished 94.5inches mirror.

Loren Shriver is an former Space Shuttle astronaut and manager of Shuttle Launch Integration, holds the Olympic Torch near the Space Shuttle Atlantis as it rests on the Launch Pad 39-A of Kennedy Space Center, July 7 in 1996. (AP Photo/Chris O'Meara)

Discovery will strive at an elevation of 380 miles, the highest that a spacecraft has ever been, meaning it can be put on its correct orbit above the Earth's shifting atmosphere. Through Hubble's 15 years of journey through space, astronomers are hoping to look back at the beginning of time , and discover the answers to one of our universe's most elusive mystery.

The Space Shuttle Discovery on top of the transporter crawler is seen approaching the service structure rotating to the right of background on the pad following leaving the building that houses the vehicle assembly located at Kennedy Space Center in Cape Canaveral, FL, June 15th (2005). (AP Photo/Peter Cosgrove)

Astro will be able to measure ultraviolet radiation that has not been detected by Hubble and, as such, offer additional targets for the telescope. This $100-million observatory will concentrate on Columbia's nine-day mission to Comet Austin, believed to be making its first journey through the solar system.

The two Hubble as well as Astro were set to launch in 1986 but was delayed by an accident on the Challenger accident.
1

## Chapter 7: Hubble Aperture Door Finally Unlocked

The cover for the lens of the Hubble Space Telescope was opened this morning following an alarming morning of problems as well. Mission Control told the Discovery astronauts that they were no longer required to save the $1.5 billion telescope.

"You've been released from Hubble support, and it's now independent to go on its own," Mission Control told the astronauts as the door to the aluminum aperture was opened.

Thank you, Sir it's a great thing"," declared Discovery Captain Loren J. Shriver.

"There's a lot of handshakes and smiles everywhere" stated Steve Hawley, who had been in charge at the time the telescope was released into space on Wednesday (April 25 (April 25, 1990). "I'll bet there's a lot of smiles and handshakes is down there too."

The success of opening the lens's cover opened the opening for the crew to return back on Earth this Sunday. The spacecraft launched Tuesday, after a delay of two weeks and removed their telescopes from the payload bay on Wednesday.

"The Hubble is open to business", Mission Control told the astronauts.

The telescope continued to have issues, however. The door was opened and two of its four position stabilizing gyroscopes became inoperative and the electronic system was put into a safety mode that is automatic that means that every movement is put to a stop.

The Hubble Space Telescope is held by the robotic arm of the spaceship Discovery approximately 380 miles higher than Earth when it was examined on April 30 1990. (AP Photo/NASA)

The team from the Goddard Space Flight Center in Maryland which is the center for the telescope, claimed that the motion of opening the doors jarred it more severely than been anticipated, and the system responded appropriately when closing the telescope.

In the early afternoon an organized process was in place to get the telescope's electronic components back on the line.

Steve Terry, a top Hubble engineer, told the press that tolerances for telescopes were set to be extremely narrow during the initial operation and was the root of numerous issues. In addition to the problems this morning and afternoon, there were two communications interruptions on Thursday that lasted for several hours.

We're cautious, which is normal the way we are,'' he stated. "We've got an extremely expensive spacecraft and we're trying not to compromise its value. ... Then we'll arrive at a point where we can know precisely what the spacecraft is going to accomplish and then set the boundaries in a sensible way."

NASA declared that at this point, the telescope is safeand steady and secure."

The spaceship Discovery that carried the telescope into orbit, was located 50 miles from the telescope, and was ready to assist should it be needed. Then, later in the day the spacecraft was scheduled to fire its engine which would put the ship on a path towards home.

The path will take that route to Discovery two miles to the left of the telescope.

A lens cover which is a 10-foot-wide "aperature door" which protects the opening of your telescope's shield of light. When shut, the telescope will be blind.

The lens cover was closed - with only the tiny wedge opening after when the $1.5 million telescope tossed onto the water by the Discovery crew last Wednesday.

When it's opened it is possible for starlight to hit it for the very first time. it will be able to go about its work of taking pictures and studying the universe with a sensitivity and an sensitivity previously unimaginable.

The door is constructed of honeycombed aluminum sheets that are covered on the outside by solar-reflecting material. Inside, the door is painted black to reflect any light that is straying.

If the door-opening command was unsuccessful and the door was not opened, those five Discovery astronauts who were deployed Hubble were ready to head for the telescope. The mission specialists Bruce McCandless and Kathryn Sullivan could have walked through space to unlock the door on Saturday in the event of a need.

If it were necessary, the rescue could delay the landing scheduled for Sunday, at Edwards Air Force Base, Calif. by one day.

The astronauts were awakened around 3:00 a.m. EDT, to soft sound that were"Kokomo." Beach Boys' song, Kokomo."

"For Max Q's instrument, this is the way it should be played and that's what Mission Control radioed to mission specialist Steve Hawley, a member of the all-astronauts band known as Max Q. The next time you'll be practicing is Sunday night in the gym, so don't be late."

Max Q is an NASA acronym that describes the moment in the spacecraft's ascent at which the pressure of the air is at its highest level.

That's similar to how we practice and I'm sure you'll agree with me,'' added Hawley. If we can be back on Sunday, I'll head out to the gym for a workout."

Astronaut Bruce McCandless II, equipped with his nitrogen-propelled manned navigation unit (MMU) spacewalks that extended to as long as 300 feet (90 meters) from the Challenger. The first non-tethered spacewalk in the history of mankind on the 9th of February 1984. (AP Photo/NASA)

The next day (May 1), NASA plans to publish the first image of a test run from the telescope, which is an open cluster of stars in the constellation Carina. The data will be significant within a one or two months.

The 12 1/2-ton, 43-foot-long telescope, which is about as big as a school bus is named in honor of Astronomer Edwin P. Hubble, who passed away in 1953. The native of Missouri, Hubble discovered in his time in the 20th century that space was expanding. His findings was the basis for the theory that the universe was created around fifteen billion years back through massive explosion.

Over the course of its 15-year-long working life, Hubble is expected to assist in determining the size and age that the Universe is. It could unravel the mysteries surrounding quasars, black holes and pulsars, and possibly even locate the stars that have planets which might provide life.

2

Nearsighted Hubble

It is believed that the Hubble Space Telescope transmitted to Earth this view of Saturn featuring an unidentified white spot, which was discovered in the latter part of Sept. by amateurs astronomers. the storm was once an earth-sized white dot to an entire circular girdle that covers the planet with ammonia clouds rising 150 miles above the ground on November 9, 1990. (AP Photo)

A focusing defect that cannot be fixed from below has been discovered within the Hubble Space Telescope and has left some instruments inoperable until repairs to the shuttle can be scheduled, NASA announced Wednesday (June 27 in 1990).).

A lot of the stunning views of the universe scientists had hoped to capture using Hubble's $1.5 billion Hubble instrument won't be realized until a planned maintenance mission for 1993 is finished according to officials.

NASA announced that it would attempt to launch the repair mission earlier.

Engineers found the flaw as they worked to refine the focal point between the primary and secondary mirrors on an orbiting telescope.

Hubble's deputy project director Jean Olivier said "there was not a single point...with an uncluttered, clear image."

The study also revealed an major flaw that could be present when the mirrors were created, Olivier said during a news conference.

Jack Rehnberg, the chief of the program for space science within Hughes Danbury Optical of Danbury Conn. The company that made the defective mirrors, suggested the problem could have been caused by human error during a long testing program when they were manufactured.

We don't know what caused it however, it's likely to be an inherently flawed, fundamentally wrong thing that was not executed correctly the way it was done," he added. It could be an error made by a human."

The mistake, he added could have been caused by equipment that was able to check the mirror's polished surface. NASA has stated that its tests proved that Hubble's primary mirror is the most smooth and largest mirror ever created.

We could have made a mistake in the construction of certain test equipment and that was not picked up even though we've conducted a lot of reviews, as well as check-ups and checks," said Rehnberg.

NASA has selected Lew Allen director of Jet Propulsion Laboratory, to be the head of a team studying the Hubble issue. Rehnberg announced that he's selected engineers from his firm to assist.

NASA Scientist Ed Weiler said the problem affects the ability to focus many instruments that are part of the orbiting telescope. This is particularly two cameras specifically designed to take images in visible light from very distant and dim stars.

He claimed that a small camera for objects, which was provided by Hubble Hubble Project from the European Space Agency, will not be able to capture photographs of any quality superior to the ground-based telescope.

A wide-field camera, that was used to shoot some of the first pictures from Hubble that were made general public, will become useless until the telescope is able to be fixed.

"There's no research we can apply to it at the moment", he stated.

Three instruments that study stars by collecting infrared and ultraviolet radiation will not be so severely damaged, Weiler said. He also said that one of the instruments, a high-speed photometer, is expected to gather "around half" of the data" that was originally designed.

The Hubble was put into an orbit that is 381 miles high by spacecraft Discovery on the 25th of April. Engineers have been trying to refine the telescope as well as test its instruments. They have reported numerous issues.

Engineers initially discovered that a cable that was not connected properly blocked the movement of the antenna. This was fixed by restricting the antenna's movement.

They later discovered the wrong computer software making the telescope be pointed at incorrect places on the sky. The software was updated.

Experts also discovered that the satellite weighing 25,500 pounds was able to produce a sound lasting for a few minutes every time it moved from sunset to sunrise and then back to its orbit. Engineers were planning to counter the vibration using a different computer program.

However, there's no solution to correct the problem of focusing on the ground, according to Weiler. Instead the telescope will be repaired, he added. NASA is planning to delay an scheduled maintenance flight for the telescope.

Weiler stated that a replacement wide-field camera which had been scheduled for Hubble could be developed using specific optics to eliminate the distortion of focusing due to the mirrors of telescopes.

We are confident that we can eliminate every speck of error that we have seen,"' said Weiler.

Hubble maintenance flights on the spacecraft were originally scheduled for 1993 1996, 1997 and 1993 and Weiler stated that the flights could be scheduled sooner if replacement instruments that will correct the focusing issue could be developed.

"We're not losing our science in any way," he added. "We are deferring science."

This color composite picture of Mars is the result of observations taken by NASA's Hubble

Space Telescope. The images were captured using an instrument called the Wide Field & Planetary Camera (WF/PC) in PC mode. Mars was at a distance of 85 million kilometers, or 53 million miles away far from Earth. The approximate .2 2 arc second resolution of this image gives Martian surface details as tiny as 50 km (31 miles) across. On the Mars photo, you can see a large blanket of blue clouds cover the icy northern polar regions in the region of Martian season at time of photograph. A large number of thin clouds located in the south hemisphere likely to be harbingers of a southern polar hoodthat could be formed when the seasons slowly change to autumn (one Martian year is 1.8 Earth years.) December 13 in 1990. (AP Photo/NASA)

This image of Jupiter was taken with The Hubble Space Telescope's broad field camera on the planets shows an oval circle on the left, along with it's the Great Red Spot just rotating out of view, to the upper right. The Hubble telescope made its first observations of Jupiter, on March 11, 1991. (AP Photo/NASA)

Olivier has been the first to test the telescope, stated that the engineers noticed patterns in their tests which exhibited the ''classic' characteristic that you'd expect from an spherical aberration in which light is not focused at an exact location however it is spread over a tiny area.

The mirrors were tested separately at the surface, according to Olivier and they did well in every test. They were not tested together on the ground as it would require an arrangement for testing which would have cost hundreds million dollars. The mirrors were ''

Prior to the launch, NASA stated that the five instruments aboard will detect objects that are 25 times more faint than the dimmest visible from Earth and increase the size of the universe that could be observed by 250 times. The objects in the outermost areas in the Universe, around fourteen billion light-years or around 82 billion trillion miles are believed to have been visible. As the universe is believed to be between 15 and twenty billion years old. A telescope will have seen objects within an hour following the so-called Big

Bang that is believed by astronomers to be the source of the universe.

First Checkout of the Hubble Telescope

The task was completed in six months rather than three months, but NASA is nearing completion of reviewing all the systems that make up the Hubble Space Telescope and is currently preparing the telescope's instruments for use in science.

The "science verification phase" is scheduled to last until next year. It will involve comparing images taken by Hubble of stars in the sky with pictures that were taken on the ground.

This allows astronomers to alter readings taken from the Hubble's five instruments just as a motorist examines his speedometer with markers on the highway and compensates for the differences.

During the phase of orbital verification that is now close to finalization it was discovered that the National Aeronautics and Space Administration discovered that one of the telescope's mirrors had been grounded according to incorrect specifications, and the spacecraft was able to produce an unsettling shake every time it moves from daylight to dark.

When we first began to plan the mission, orbital verification was to take about 3 months," explained Jim Elliott, spokesman for the Goddard Space Flight Center. Because of the issues with the mirror's aberration as well as the jitter, we'ven't been able to complete all we had hoped for within that time."

J. Keith Kalinowski, director of the telescope's Science Management Office at Goddard Kalinowski, told the Goddard News that the whole calibration process was anticipated to take about nine months. In reality, he said it could take between 15 to 18 days. The Hubble was taken off the spaceship Discovery in April. Despite its issues however, Hubble has produced better images than the ones taken by the ground. One of them is the clearest images yet of Pluto the most distant planet from the sun's solar system, and also its moon Charon.

Hubble also captured Saturn with greater depth than was previously possible at a distance, however it falls it was short of the clarity that was achieved by the approaching Voyager spacecraft.

A galaxy that was only observed as a fuzzy ball with telescopes was discovered to be so full of stars it could have a black hole at its central point. The galaxy is around 40 million light years from us. It has around 30,000 stars within an area of about 9 light years.

A new European Space Agency Faint Object Camera image of Pluto and its satellite Charon is displayed in the upper right-hand corner of this image. To the left of it is a picture of Pluto and Charon captured by the earth's telescope. This Hubble photograph is first long-duration image of a moving target on October 1 90. (AP Photo)

We've lost a lot of sensitivities and the ability to recognize extremely faint objects, according to Kalinowski. In general, if I could detect an object with some luminosity at a particular distance within a certain exposure time, I might only be able to see the identical object when it's half the distance from me."

The instrument that is most affected by the mirror issue was that of the Wide Field Planetary Camera, that was supposed to detect light that came from the time of the universe's beginning. The current plans call for an on-board shuttle crew upgrade the cameras in the year 1993 by one equipped with lenses that are correct.

A different plan under consideration will restore three other instruments to their original sensitivity including corrective lenses into one of the bays for instruments.

Kalinowski said that a feasibility report is expected to be completed within about a month and said it looks promising."

Space News, an industry publication, has revealed that the Smartstar"'' is a $7,000 device designed in collaboration with University of Wisconsin. University of Wisconsin. The estimates of the cost for the project are between $20 million and $60,000 or higher as per Space News.

3

Hubble Servicing Missions


Astronaut Andrew Feustel, left, is navigating near his Hubble Space Telescope at the far end of the remote manipulation arm, which is controlled from the inside of Atlantis the crew cabin. Astronaut John Grunsfeld, right, communicates with his crewmate from a mere couple of feet from. Feustel and Grunsfeld were working on maintenance work on the huge telescope, which was locked in the cargo area of the spacecraft, May 16 09, 2009. (AP Photo/NASA)

It's four years behind and will cost additional twenty million. However, by 1994, the Hubble Space Telescope is expected to function as it was designed to.

The National Aeronautics and Space Administration is quietly working to rectify the major embarrassment that occurred last year which was the finding that the lens of the $1.5 billion telescope was incorrectly positioned which led to it being nearsighted.

While no official declaration has yet been issued however, there is a NASA contractor is working on corrective lenses to be put in place by spacewalking astronauts later in 1993, or most likely, in 1994. A replacement for solar arrays that cause jitter is being developed.

"The new Hubble will accomplish everything the original Hubble could ever do as well, according to Charles Pellerin, director of NASA's astrophysics division, stated in an interview. The new Hubble will, he stated, a few very minor hiccups however, it will achieve the full 97 percent of its scientific potential."

The promise was stunning. "The Hubble Space Telescope will gather light from objects that are 25 times more faint than the current telescopes" the blurb prior to the launch announced. The telescope will allow you to look further into space and see objects 10 times the resolution of the top Earth-based instruments. The optics on the telescope are the most advanced ever made."

But a "spherical distortion of the lens blocks the telescope from observing the dim light, which was formed millions of years ago, in the time when our universe was just beginning to grow.

Director of the flight Robert Castle, lower right has laptop computers to assist his work at NASA's Mission Control during one of five space walks that were conducted to maintain the Hubble Space Telescope in December 1993. (AP Photo/NASA)

The Hubble's shortcomings have been exacerbated by comics and NASA critics

However, Pellerin stated that it's still the most powerful telescope in the world in all of the world, bar none. It is the best telescope in the world. He added that the telescope has been making discoveries every single week.

Its feats that are all within relatively short distances, include images of a spectacular storm on Saturn and observing the star as "incredibly rich" with platinum; locating an astrophysical jet stream of gas that is moving across hundreds of thousands of kilometers per hour; and proving that certain light points that appear fuzzy actually contain complicated stars.

NASA has not yet announced an end-to-end decision regarding the Hubble repair, but the production is underway according to Pellerin. Plans are being developed for training two astronauts to complete three spacewalks each.

On the first day they will replace Hubble's solar arrays , which provide the telescope with electricity to instruments and transmitters. The current arrays, which are provided through the European Space Agency, have two issues.

Each 90 minute period, as the telescope is moved from darkness to the Earth to the light side the sudden temperature shift will cause the rods that frame these solar panels to be bent.

At the point of the telescope, the motion is around 11 inches in length, according to Pellerin explained. The telescope shakes. The telescope shakes.

The telescope is moving from the 221-degree Fahrenheit (the boiling point of water) the sunlit side of Earth to the cold dark side, which is 180 degrees below freezing The solar blankets expand. The movement is thought to be compensated by rolling however one roller is stuck and causes a new instance of shaking every 90 minutes.

Both issues will be addressed using the new arrays Pellerin stated.

The next day, astronauts will take over the Hubble's Wide-Field Planetary camera. They will also install the 700-pound COSTAR box. COSTAR. COSTAR is a device that allows you to install mirrors that are the equivalent to postage stamps the direction of two spectrographs, and a camera that is faint.

COSTAR is a phone booth-like device. telephone booth, will be able to fit in the same space on the Hubble which is currently used by a device that measures the intensity of light. The acronym refers to Corrective Optics. State Telescope Axial Replacement.

The camera that is wide-field collects images of planets as well as other stars in the solar system as well as distant objects like galaxies and quasars , which contain clues about the beginning of the history in the Universe. The camera was scheduled to be replaced in 1993, but it was not scheduled for replacement. The new model was modified to eliminate the mirror flaws.

It is also important to note that the repair effort had been planned from the beginning will help keep the additional cost of COSTAR and other repairs to around $20 million, according to Pellerin.

In the third day of spacewalks the new set of Gyros will be put in the Hubble that has 4 sets that are two.

In the last few months one of the Gyros started making an electronic noise , and was turned off.

Gyros are the devices that notify controllers of what actions the telescope is making.

Pellerin confirmed that the repairs will result in the Hubble nearly complete.

The surgery is now to get the patient back. the doctor said "not to become a jogger instead, but to become an Olympic track star."

NASA's Record Five Spacewalks On Hubble Mission

NASA's record five spacewalks in order to fix the Hubble Space Telescope, led by astronauts Story Musgrave, Jeffrey Hoffman, Kathryn Thornton and Tom Akers:

First Spacewalk, Dec. 5

Musgrave and Hoffman take 7 hours, 54 minutes to set up 2 pairs of Gyroscopes two electronic units for gyroscopes and eight fuse.

Second Spacewalk December. 6

Thornton and Akers spend 6 hours and 36 minutes, removing an abrasive Hubble solar array, and putting up two new solar panels on the telescope.

Third Spacewalk Dec. 7

Musgrave and Hoffman spend 6 hours and 47 minutes to set up the latest wide-field planetary camera as well as the two magnetometers.

Fourth Spacewalk December. 8

Thornton and Akers spend 6 hours and 50 minutes to install a huge set of mirrors that correct, as well as additional memory for computers.

Fifth Spacewalk Dec. 9

Musgrave and Hoffman spend 7 hours and 21 minutes, replacing the equipment for the telescope's solar arrays, and also to put in a switch for the ultraviolet detector.

"Costar," a detail of which can be seen here, was created to correct focus within Hubble's Hubble Space Telescope, and is one of two brand new objects from Hubble that are on display in "Moving Beyond Earth,"" an exhibition in the National Air and Space Museum and is displayed inside the Museum in Washington on November. 18 19th, 2009. The new gallery has ample space to include more artifacts over the next years as NASA will end its Space Shuttle program. Following this mission is completed, there will be only five missions are left. (AP Photo/Jacquelyn Martin)

U.S. President Bill Clinton with Vice-President Al Gore, talk by telephone via to the Oval Office in Washington, DC. together with Hubble Mission astronauts to congratulate their team for the successful completion of their work on Hubble Space Telescope. Hubble Space Telescope. The Hubble Mission astronauts are displayed on a monitor on television at left, in their bi-directional long distance call on 10 December 1993. (AP Photo/Greg Gibson)

Hubble Telescope Fixed, Taking Sharp Pictures

NASA officials have declared they had declared the Hubble Space Telescope repaired on Thursday (January 13 1994) and announced that the orbiting observatory is now able to snap the most clear, sharp photos ever of ancient and distant stars.

The telescope has been repaired above our expectations and we're very happy with it,'' declared Ed Weiler, NASA's Hubble program scientist. NASA has provided an instrument that has fulfilled every promise we made prior to launch."

Hubble $1.6 billion Hubble was launched in the year 1990 with a faulty magnifying mirror, which resulted in a blurred view that of all the stars. The telescope was damaged in December 1993. in the most complicated repair mission in space history an astronaut from the spaceship fitted corrective optics and a brand new camera on the telescope.

The first photos taken from the new instruments show sharp, precise individual light points of galaxies as far as fifty million light-years far away. The fuzzy smears of light that were produced by the older Hubble. Stars that were once were obscured by a milky blur are now clearly visible.

"I felt it was the right thing to do and I knew it was right, astronaut Story Musgrave, who led the repair mission, told reporters in Houston. But, you know, it was so difficult."

James H. Crocker of the Space Telescope Science Institute in Baltimore explained that the view of the telescope fixed is so precise that , from Washington D.C. it could record the blink of a firefly from Tokyo approximately 8500 miles away.

If there were another firefly that was just 10 feet away from the one said Crocker the

Hubble will be able discern there were two lights bugs, not just one.

Hubble Hubble could read the headlines of newspapers in 12 miles distance, Hubble said.

The images that have been corrected are as flawless as the engineering can achieve and as perfect as physical laws allow for it," explained Crocker.

Weiler explained that the older Hubble with its blurry image, was able to collect crucial scientifically-based views of stars that were of three to four billion light years from. With the latest optics and the new camera, the Hubble will take pictures of stars between 10 and eleven billion light years distant, which is close to the theoretical limit in the cosmic sphere.

NASA Director of Science and Technology, Associate Administrator Science Mission Directorate Dr. Edward J. Weiler discusses new images released of the Hubble Space Telescope at NASA Headquarters in Washington. The images are taken from four of the telescope's six instruments for science operation September 9th September 2009. (AP Photo/NASA, Bill Ingalls)

With a ground-based telescope, he explained that galaxies that are appear far away are only observed as fuzzy balls."

David Leckrone, senior project scientist at the Goddard Space Flight Center, revealed that the new Hubble will begin conducting serious research into the universe this week, though testing some of the instruments on the spacecraft may take time. Spectrographs, which study the light spectrum at different wavelengths have yet to be fully tested, however they operate on the same optical systems that are used by Hubble cameras. Hubble cameras.

Weiler stated that the rebuilt Hubble is able to answer fundamental questions about the universe.

There are several examples:

• What's the dimension and time span of the universe? And how rapid are they expanding?

* Does the universe appear to be open or closed? If so expanding forever or, eventually cease expanding, and fall apart?

Are black holes really exist? Hubble, the fixed Hubble is able measure the velocity of stars that are pulled towards centers of black hole.

With these measurements that, according to Weiler"We'll be able say there exists the black hole."

What is the process by which matter develops? Every atom is formed by stars, and scientists are keen to understand more about the process.

NASA honored its Hubble repair by holding three news conferences that were a success which included the NASA administrator and an White House advisor and a U.S. senator.

In 1990 the year 1990, when NASA admitted that Hubble was not working it was announced at a press conference that was that was attended by only second-tier NASA engineers and senior executives.

Daniel S. Goldin, the administrator of NASA, said that the success of the repairs "solidifies that we're an agency that can be done."

John H. Gibbons, vice president of technology and science, said the intricate repair of Hubble was "one of the most challenging projects NASA has done." The agency, he noted was able to complete its mission with speed and on budget."

In recent times, NASA has been plagued by issues. Apart from Hubble's poor optics the space agency has lost an orbiter that was on its way to Mars Another space probe has been hampered by a jammed antennae and the agency is plagued by cost overruns and is eroding support from Congress for the space station.

Sen. Barbara Mikulski, D-Md. she said that the images taken by Hubble are tangible proof that not only is Hubble been repaired however, NASA is on the path to redressing the culture that led to these issues."

2. Mission 2 (STS-82): Space Shuttle Discovery Discovery captures Hubble Telescope for Upgrades

Over Mexico's southwest coast An astronaut was able to reach out with the crane from spaceship Discovery and shrewdly pulled from the Hubble telescope off its isolated wandering through space.

NASA Astronaut Steven Hawley then swung the 43-foot-tall telescope across the spacecraft's open cargo area and slowly dropped it down onto a platform to be subject to four consecutive days of swapping components to boost its already remarkable performance.

"Gee I'd love it if I could have seen the smile that the face of Dr. Stevie's expression," declared the Commander Kenneth Bowersox. "Looks like he's shaken the hands of an old acquaintance."

Hawley was the one in charge as well, when the Hubble was put on its own in the year 1990. Hawley will be the operator of the telescope's arm when it will be released next the 18th of February (February 18, 2018).).

Everything went according to the script, though 17 minutes behind schedule. The Discovery's 31st orbit Bowersox directed the shuttle to the telescope, which was 368 miles higher than Earth.

"It was a bit slow in the final moments", Bowersox said. Bowersox laughed. "We were just trying to make our rendezvous last longer. It was a lot of enjoyable."

Before taking the picture, Bowersox answered an important query that the solar panels of the telescope were unharmed by their lengthy exposure to harsh atmosphere of space.

Hubble Space Telescope Hubble Space Telescope is a part of the Space shuttle

Discovery payload bay following its capture during one of the four first spacewalks designed to improve the technology of the billion-dollar Hubble launch in the year 1990. The new instruments include a spectrograph that has two-dimensional detectors as well as an infrared camera are expected to be 30-40 times more effective and powerful than the older ones and let astronomers look through the sky almost from the beginning of time January 1997. (AP Photo/ NASA)

The program will begin tonight and continue each night until Sunday, two pairs of spacewalkers will explore the cargo bay to replace 11 telescope parts of 1970s design with the latest components.

The Hubble is close to the halfway point of its 15-year lifespan However, no emergency has led to the modernization. The service call was different from one made in December 1993 when the corrective optics were installed in a hurry to improve the telescope's blurred vision.

NASA determined that the catch was on orbit 37,130 and the telescope had recorded just under one billion miles.

The result from all the star-gazing activities have been photos looking back further back in time and further away like no other. There hasn't been any revolution in studying our universe before the discovery of the telescope over 400 years ago.

"There is five times as many galaxies as we imagined," stated Ed Weiler, NASA's chief Hubble scientist. In the wake of an astonishing image compiled from 342 images taken in 1995 Weiler stated that "estimates ranged to 10 billion (galaxies) and 50 billion."

Free of distortions caused by the atmosphere, Hubble has photographed stars in the midst of a fiery blaze of death, as well as towering clouds of cool molecules of hydrogen as well as dust from which stars form. The telescope observed a meteor smash into Jupiter as well as a collision of two galaxies that were traveling at 1 million mph. It also it revealed the unimaginable the surface of Pluto.

Since the launch on Tuesday (February 11) launch, Discovery has played catch-up with the telescope, slowly closing the gap of 7,500 miles through a series rocket explosions until they were together.

In the hours before midnight tonight, Col. Mark Lee and Steve Smith will put on space suits and put in two $100 million instruments that are each about as big as a telephone booth A near-infrared camera as well as an imaging spectrograph with two dimensions.

Astronaut Steven Smith works at the end of the spacecraft's remote manipulator as they perform maintenance on the Hubble Space Telescope during an orbital walk. The background is a part of Australia around the Earth's curvature on February 15, 1997. (AP Photo/NASA, File)
The spacewalking crew of the other, Gregory Harbaugh and Joe Tanner will leave on Friday evening. All in all eleven major components are scheduled to be put in place.
The Fifth Spacewalk

In the fifth spacewalk, which should have been the last spacewalk of the servicing mission on Monday night (February 17th 1997), Lee and Smith utilized bits of foil, clips, wire, twists and parachute cords to restore Hubble's melting insulation.

The spacewalkers draped quilt-like patches on fractures in Hubble's reflective insulation, which was damaged by sunlight exposure over seven years of the orbit. They pinned the six 9-inch by 16-inch pieces of material onto knobs and rails on the telescope.

In the area in which the insulation had broken but not yet completely ripped in the area, the Discovery crew extended two wires to keep the material from splitting.

It was an exhausting job. Two times, Lee cursed. Spacewalkers were required to work with the help of their helmets. Most of the two and a half hours was carried out in the dark of space.

Mission Control added the spacewalk to Discovery's flight of the shuttle so that Lee Smith and Lee Smith could complete the insulation work that was started by two of their colleagues on the night prior.

The astronauts discovered the damages in the last week when they installed state-of-the-art technology that allows the telescope to see further into the universe.

With the same creativity that was used in Apollo 13, the crew made the patches Monday morning as Gregory Harbaugh and

Joe Tanner put in the final Hubble replacement components, and also repaired a few of them also.

In 375 miles of work higher than Earth, Harbaugh and Tanner have filled in two gaps close to on the summit of their 43 foot telescope by using Teflon coated pieces that measure three feet long and one foot wide. They affix the blankets, used to repair potential pinholes, to knobs as well as rails with strings and wire.

The task of putting the patches made by hand over the lower compartments of electronics was deemed to be more difficult and even more crucial. The astronauts pulled the material that was intended for this kind of an issue in the cargo area.

NASA managers were pleased at how well the first repair were completed. "It was a pleasant feeling," declared Mike Weiss, a Hubble service manager.

The repairs were not as important as the ones made during Apollo 13's failed moon mission in the year 1970. Three astronauts brought their life back applying tape and cardboard covers ripped off of their flight manuals in order to repair the spacecraft's mechanism for purifying the atmosphere free of carbon dioxide.

Hubble actually may have survived to the next scheduled service call in late 1999 without the insulation repair, NASA payload manager Kenneth Ledbetter stated. The worry was that the damaged cover could cause the delicate electronics inside this $2 billion space telescope to get too hot and eventually fail.

"It is something that we thought was sensible to donot necessarily essential, but it was prudent to do, and so we did it," Ledbetter stated.

Harbaugh Tanner and Harbaugh Tanner were extremely proud of their accomplishments. They worked for 1 half hours securing two blankets before making sure they were adjusted just right.

The Hubble Space Telescope is shown following its release from the space shuttle Discovery, February 19, 1997. (AP Photo/NASA, File)

"What do you think?" Harbaugh asked, backing away.

"Like it. Looks good from here," Tanner replied.

Mission Control put it this way: "A masterpiece."

NASA plans a more permanent fix during the next service call in three years. The astronauts snipped off a piece of the damaged insulation to bring back home for analysis.

The astronauts are scheduled to release the Hubble on Wednesday (February 19), from the shuttle's cargo bay, where it has been anchored since last week. Discovery is scheduled to return to Florida on Friday.

Mission 3A (STS-103): Space Shuttle Discovery
NASA Clears Discovery

After a weekend of repairs, NASA today cleared space shuttle Discovery for liftoff Thursday (December 16), to the Hubble Space Telescope.

Pilot Scott J. Kelly, left, holds a model of the Hubble Space Telescope during a televised interview from Discovery's flight deck with various television networks. Joining Kelly from left to right are Jean-Francois Clervoy, Michael Foale, and commander Curtis L. Brown with radio, December 20, 1999. (AP Photo/NASA TV)

Technicians successfully replaced a dented fuel line that had delayed the repair mission by nearly a week. The flight already was running two months late because of shuttle wiring repairs.

This morning, shuttle managers gave the go-ahead for the launch countdown to begin early Tuesday. Liftoff is scheduled for 9:18 p.m. Thursday.

NASA has only three chances to launch Discovery this month: Thursday, Friday and Saturday. If the shuttle isn't up by then, the mission will be put off until January to avoid having the shuttle fly over New Year's, which could pose Y2K problems.

Space shuttle Endeavour headed to NASA's other launch pad today for a late January launch on an Earth-mapping mission.

Astronomers are eager for Discovery and its seven astronauts to take off as soon as possible. Hubble is shut down because of a failed pointing system and cannot resume astronomical observations until the crew replaces gyroscopes and other parts.

Mission 3B (STS-109): Space Shuttle Columbia Astronauts Attach Wing to Telescope

A fresh pair of spacewalkers attached a second powerful 25-foot solar wing to the Hubble Space Telescope on Tuesday (March 5, 2002), with speed and apparent ease.

It was the second of five excursions planned this week by space shuttle Columbia's telescope-repair team.

"Beautiful day for a spacewalk," astronaut James Newman observed as he left the shuttle. "Incredible," said Michael Massimino, a first-time spacewalker.

The men quickly moved through their chores 360 miles up, removing and stowing the old solar wing, and unpacking and lifting the new one into position. It mimicked Monday's work by two other astronauts, who installed the first new solar wing on Hubble.

Massimino tightly gripped the new 640-pound wing and held it steady as he rode the shuttle robot arm to the attach point on the telescope. "Lean back a little bit," Newman told him.

"Take a deep breath and relax. You did a great job."

Less than an hour later, Massimino unfolded the hinged wing like the cover of a storybook, opening it at a cautious rate of one degree a second. "Ready? Here it comes," he said.

With that completed, the spacewalkers began replacing an unreliable steering mechanism.

One of Hubble's four reaction wheels briefly malfunctioned in November. Although the wheel has since been working fine, engineers concluded the failure was likely to occur again.

At least three of the 100-pound, cylindrical wheels are needed to aim the observatory at intended targets. Four speed things up.

Both the wheel and wing passed their initial tests. The work went so well, the spacewalkers had extra time to do some minor latch and insulation repairs.

Astronaut John Grunsfeld, left, traverses along the longerons of the Space Shuttle Columbia while astronaut Richard Linnehan uses the Remote Manipulator System's robotic arm to move around during spacewalk. Grunsfeld and Linnehan spent seven hours outside installing a new solar wing on the Hubble Space Telescope, March 4, 2002. (AP Photo/NASA)

Hubble's new $19 million set of wings, covered with ultra-efficient solar cells, should boost electrical output by more than 20 percent. That will be especially useful once an advanced camera is added later this week and two more science instruments are launched in two years.

The telescope's old solar wings had become damaged by eight years of wear and tear.

The new rigid wings also will reduce the rate at which the 43-foot, 24,500-pound Hubble gradually sinks in orbit.

Space shuttles periodically give a little boost to the telescope. But NASA plans to stop servicing Hubble in 2004 and to bring it back for museum display in 2010.

"We really hope they'll give us a nice, long science lifetime on Hubble Space Telescope," said Anne Kinney, NASA's astronomy director.

The astronauts also did some advance work for the next spacewalk, considered the most complicated one of the mission.

Massimino loosened bolts on doors leading to Hubble's batteries and power-control unit.

That will save John Grunsfeld and Richard Linnehan precious minutes when they disconnect the batteries and attempt to replace the control unit on Wednesday, March 6.

The observatory will be completely turned off for the repair, for the first time ever in orbit.

Repaired Hubble Released Into Orbit

Their mission accomplished, space shuttle Columbia's astronauts released a more energetic and scientifically potent Hubble Space Telescope into orbit Saturday (March 9, 2002), after five days of repairs.

"Good luck, Mr. Hubble," astronaut John Grunsfeld called out.

Shuttle crane operator Nancy Currie set Hubble free as the two spacecraft zoomed 360 miles above the Atlantic. She had used Columbia's robot arm last weekend to capture the 43-foot, 24,500-pound telescope and anchor it in the cargo bay.

Columbia slowly backed away from Hubble, giving the world its last close-up look at the telescope until astronauts return for another overhaul in two years.

Astronaut John Grunsfeld signals to a crew mate inside the cabin while working in the payload bay of space shuttle Columbia during a spacewalk to install a new solar wing on the Hubble Space Telescope, March 4, 2002. (AP Photo/NASA, File)

A beautiful view of Mr. Hubble, the telescope, over the Earth's horizon, ready to go and make new discoveries," said Grunsfeld, the chief telescope repairman. "We bid Hubble well on its new journey with its new tools to explore the universe. Good luck, Mr. Hubble."

Replied Mission Control: "And we wish it well from down here as well."

Over the past week, two teams of spacewalking astronauts outfitted Hubble with smaller but more powerful solar wings, a more robust central power controller, a pointing mechanism, an advanced camera for peering deeper than ever into the universe, and a super-cold refrigerator for revitalizing a disabled infrared camera.

All the new components passed initial testing. It will be at least a month, however, before NASA knows whether the experimental cooling system was able to get the infrared camera working.

Columbia's crew of seven, meanwhile, is due back on Earth on Tuesday.

More nerve-racking than any single repair was the complete shutdown of the $2 billion-plus telescope for the replacement of the central power controller. It was the first time in Hubble's 12 years in orbit that all its systems were turned off. The shutdown lasted 4 1/2 hours; to everyone's relief, everything came back on when power was restored.

NASA considered this service call — the fourth — the most complicated one yet. Wednesday's operation, to replace the power controller, was likened to heart transplant surgery.

"Many people on this mission, privately, didn't think that we would be able to accomplish everything that we had set out in our plan,'' Hubble program manager Preston Burch said after the fifth and final spacewalk ended Friday.

Columbia's astronauts set a spacewalking record for a single shuttle mission: 35 hours and 55 minutes. It surpassed, by 29 minutes, the 1993 record held by the first Hubble repair team.

One last servicing mission is planned in 2004. Then Hubble will keep surveying the heavens until 2010, when a shuttle is slated to retrieve it and bring it back to Earth. NASA hopes to display the telescope in the Smithsonian Institution's National Air and Space Museum.

Hubble's successor, the Next Generation Space Telescope, is scheduled for launch in 2009.

Mission 4 (STS-109): Space Shuttle Atlantis Spacewalkers Use Brute Force for Hubble Repair

Spacewalkers' specially designed tools couldn't dislodge a balky bolt interfering with repairs at the Hubble Space Telescope. So they took an approach more familiar to people puttering around down on Earth: brute force.

And it worked. But it set spacewalkers so far behind that they couldn't get all their tasks done.

Atlantis astronaut Michael Massimino couldn't remove a 1.25-inch (3.2-centimeter) long bolt attaching a hand rail to the outside of a scientific instrument he needed to fix. The rail had to be removed or at least bent out of the way.

That was only the beginning of a hard-luck day. The balky bolt and other tiny problems put spacewalkers so far behind schedule that they had to abandon the second part of their spacewalk: replacing some worn insulation on the telescope.

NASA, which prides itself on being prepared, had not anticipated a bolt problem while removing the 1.5-foot (0.5-meter) long hand rail, said lead flight controller Tony Ceccacci.

The crew of space shuttle Atlantis from left, mission specialists Michael Good, Michael Massimino, Andrew Feustel, John Grunsfeld and Megan McArthur, pilot Gregory Johnson and commander Scott Altman leave the Operations and Checkout building during a dress rehearsal for their upcoming mission at Kennedy Space Center in Cape Canaveral, FL, September 24, 2008. (AP Photo/John Raoux)

Astronomers, whose nerves were tried by the spacewalk, were still happy because it was the second straight resurrection of a much-used but dead scientific device.

"The science capabilities we've been given today are fabulous," Jennifer Wiseman, NASA's chief of stellar astrophysics said at a late Sunday (May 17, 2009), news conference.

"It's almost like starting with a brand-new observatory."

The marathon spacewalk by Massimino and Michael Good took so long — just more than eight hours — that it was the sixth longest U.S. spacewalk and a few minutes longer than the one Friday (May 15).

When several tries with different expensive tools couldn't remove the stripped-out bolt, Mission Control in Houston told Massimino to go for the less precise yank.

Technicians prepare hardware that will be carried on space shuttle Atlantis to NASA's Hubble space telescope at Kennedy Space Center in Cape Canaveral, FL. Five spacewalks are planned during the 11 day mission to install new instruments and repair and replace other systems in the telescope, September 10, 2008. (AP Photo/John Raoux)

At Goddard Space Flight Center in Maryland, engineers twice tested that pull on a mock-up before Massimino was told to use his muscles.

The space shuttle Atlantis, aboard a NASA 747, departs the NASA Dryden Flight Research

Center at Edwards Air Force Base, CA, enroute to the Kennedy Space Center in Florida, beginning the last leg of STS-125, its mission to repair the Hubble space telescope, June 1, 2009. (AP Photo/Reed Saxon)

"You hope you don't get to the point where you just close your eyes and pull and hope nothing (bad) happens," said James Cooper, the Goddard mechanical systems manager for the repair mission. "But we had run out of other options."

Astronauts were careful to tape pieces so they wouldn't fly away and become potential missiles.

"This is like tying branches together in Boy Scouts," Good said.

Since Atlantis was out of video contact 350 miles (563 kilometers) above Earth, controllers in Houston could only listen as Massimino took a breath and pulled.

After a second of silence, Massimino calmly said: "disposal bag, please."

After nearly two hours of work on the balky bolt, astronauts went back to the plan to bring the Space Telescope Imaging Spectrograph back from the dead.

Early test results show the spectrograph, disabled by a power failure five years ago, was brought back to life. When further tests started, a glitch popped up, but NASA officials were confident the device would be fine.

Three of the four Hubble spacewalks so far have been delayed by niggling problems, like stubborn bolts and objects that wouldn't fit. A fifth and final spacewalk is set for Monday.

Massimino's run of bad luck continued. While trying to install a special plate to remove 111 tiny screws that held the instrument cover in place, a tool's battery died. It took more than half an hour for him to go back to the shuttle, swap out batteries and recharge his oxygen supply.

By the time Massimino replaced the internal electronics power supply card in the spectrograph, it was just about the originally scheduled time for the end of the spacewalk. And more than 90 minutes of clean-up and close-out work remained.

So spacewalk coordinators on the ground decided that the second part of Sunday's task, the insulation, had to be put off until Monday, if possible.

"We're very proud of you," Atlantis astronaut John Grunsfeld told the weary spacewalkers.
Grunsfeld and Andrew Feustel will not only pick up some of their work Monday, but they will be the last people to touch the 19-year-old observatory.

Space shuttle Atlantis sits on the launch pad at Kennedy Space Center in Cape Canaveral, FL, March 31, 2009. (AP Photo/Peter Cosgrove, File)

On Tuesday, Atlantis will release Hubble, which NASA hopes will keep operating for another five to 10 years, before it is steered to a watery grave.

On Saturday, two other astronauts revived Hubble's survey camera. Early Sunday, Mission Control told the crew two of the three science channels on the repaired camera were working again.

When NASA planned this mission, officials said it would be a success if either of the two dead instruments could be revived. With Saturday's camera remedy, fixing the spectrograph is a bonus.

The light-separating spectrograph has helped find black holes and examine the atmosphere of planets outside our solar system.

All the work may not get done Monday, but at least part will be attempted, Mission Control said.

## Chapter 8: Five Key Discoveries
Before Galaxies, When Stars Were Young

The Hubble Space Telescope after its release from the Space Shuttle Atlantis as the two spacecraft continue their relative separation after having been linked together for the better part of a week. During the week five spacewalks were performed to complete the final servicing mission for the orbital observatory, May 19, 2009. (AP Photo/NASA)

Key Discovery 1: Evolution of Galaxies

The young cosmos was filled with much smaller and more irregularly shaped galaxies, probably the building blocks of spiral and elliptical galaxies now seen in the older universe.

A 12-billion-year look back in time has given astronomers a glimpse of the universe when it was barren of galaxies and of an era when stars were undergoing a "baby boom" of formation.

A long exposure picture by the Hubble Space Telescope in 1996 focused on an almost blank speck of sky has revealed a vast population of unseen galaxies and looked beyond to a time when the universe appeared blank and empty, astronomers said Thursday (May 8, 1997), at Baltimore's Space Telescope Science Institute.

Several hundred never-before-seen galaxies are visible in this "deepest-ever" view of the universe, called the Hubble Deep Field (HDF), made with NASA's Hubble Space Telescope. Besides the classical spiral and elliptical shaped galaxies, there is a variety of other galaxy shapes and colors that are important clues to understanding the evolution of the universe. Some of the galaxies may have formed less than one billion years after the "Big Bang." Representing a narrow "keyhole" view all the way to the visible horizon of the universe, the HDF image covers a speck of sky 1/30th the diameter of the full Moon (about 25% of the entire HDF is shown here), January 15, 1996. (AP Photo)

"We have now looked through to the other side, to a time before galaxy formation," said Richard Ellis of the Institute of Astronomy, Cambridge, England. "We're seeing a black sky before galaxies."

"This is the best picture we have yet of the early universe," said Mario Livio of the Space Telescope Science Institute. He said the images captured by Hubble over an exposure of many hours show the farthest point ever captured in the visible wavelength.

The picture, called the Hubble Deep Field Survey, covered a minute wedge of the sky, using that tiny area as a keyhole to a much larger expanse. The astronomers purposely selected a view that, to conventional telescopes, seemed almost empty. But with the long exposure, the space telescope gathered enough light to reveal more than 3,000 galaxies, each with billions of stars.

In such astronomy images, distance equals time. The farther away an object is, the older it is and the nearer it is to the birth of the universe.

The face of a star called Betelgeuse as seen with the Hubble Deep Field (HDF), made with NASA's Hubble Space Telescope. This photo, part of the deepest view ever made of the universe, was presented to the 178th meeting of the American Astronomical Society in San

Antonio, TX, January 15, 1996. (AP Photo/NASA)

Livio said the view goes back to a time when the universe was only about 10 percent of its present age. The exact age of the universe is controversial, with estimates ranging from 13 billion to more than 15 billion years.

Roger Blandforth of the California Institute of Technology said the deep field survey has so much information that astronomers will be studying it for years. The image is like a visual core sample of the sky — it includes shallow, nearby stars along with very deep, very distant objects.

"Right now, it is like unripe fruit of an incomplete crop," said Blandforth. "There are an awful lot of galaxies out there. It's much more dense than I expected."

This composite photo by the Hubble Space Telescope, a long-duration exposure and the deepest-ever view of the universe, looks back to the edge of the big bang and shows a chaotic scramble of odd galaxies smashing into each other and re-forming in bizarre shapes. The galaxies in this panel were plucked from a harvest of nearly 10,000 galaxies in the Ultra Deep Field, the deepest visible-light image of the cosmos. The image required 800 exposures taken over the course of 400 Hubble orbits around Earth. The total amount of exposure time was 11.3 days, taken between September 24, 2003 and January 16, 2004. (AP Photo/NASA)

The deep field images have been circulated to astronomers worldwide and the analysis is only about half done, he said.

Some of the most distant galaxies are seen as only half-formed, messy groups of stars that have yet to even out into the sleek and graceful forms seen in many of the closer galaxies.

Livio said it is as if the Hubble is looking at a construction site for galaxies, a place where they are still being built, and the site is littered with the detritus of stars and gas and dust.

"The most distant galaxies appear to have been disturbed, as if there were violent collisions," he said. "Those first galaxies don't appear to be nice and neat."

Among the discoveries is what the astronomers are calling a "baby boom" in star formation. Stars, formed from dust and gas that compress by gravity until nuclear fires are lighted, were sparse in the farthest point of the Hubble Deep Field. But closer, nearer the foreground of the view, when the universe was only about 20 percent of its present age, stars were being created at a furious rate.

After that, said Livio, the rate of star birth slowed.

"The star baby boom peaked, and it has been declining since," he said.

The Space Telescope Science Institute is operated by Johns Hopkins University and has the primary science role in the operation of NASA's Hubble Space Telescope.

New Theory on Big Bang

January 9, 2002

By Paul Recer

An outburst of star formation ended a half billion years of utter darkness following the Big Bang, the theoretical start of the universe, according to a study that challenges old ideas about the birth of the first stars.

An analysis of very faint galaxies in the deepest view of the universe ever captured by a telescope suggests there was an eruption of stars that burst to life and pierced the blackness very early in the 15 billion-year history of the universe.

The study, by Kenneth M. Lanzetta of the State University of New York at Stony Brook, challenges the long held belief that star formation started slowly after the Big Bang and didn't peak until some 5 billion years later.

"Star formation took place early and very rapidly," Lanzetta said Tuesday (January 8, 2002) at a National Aeronautics and Space Administration news conference. "Star formation was 10 times higher in the distant early universe than it is today."

These images show the sky in the region of the Hubble Ultra-Deep field taken with the new Wide Field Camera 3 Infra-red imager (WFC3/IR) on HST. It is the deepest image of the sky ever obtained in the near-infrared, left. Right is the image enhanced showing the galaxy that existed 480 million years after the Big Bang and the position in the Hubble Ultra Deep Field (HUDF) where it was found, undated. (AP Photo/NASA)

Lanzetta's conclusions are based on an analysis of what is called a deep field study by the Hubble Space Telescope. To capture the faintest and most distant images possible, the Hubble focused on an ordinary bit of sky for more than 14 days, taking a picture of every object within a small, deep slice of the heavens. The resulting images are faint, fuzzy bits of light from galaxies near and far, including some more than 14 billion light-years away, said Lanzetta.

The surprise was that the farther back the telescope looked, the greater the star-forming activity was.

"Star formation continued to increase to the very earliest point that we could see," said Lanzetta. "We are seeing close to the first burst of star formation."

Bruce Margon of the Space Telescope Science Institute in Baltimore said Lanzetta's conclusions are a "surprising result" that will need to be confirmed by other studies.

"This suggests that the great burst of star formation was at the beginning of the universe," said Margon, noting that, in effect: "The finale came first."

"If this can be verified, it will dramatically change our understanding of the universe,'' said Anne Kinney, director of the astronomy and physics division at NASA.

In his study, Lanzetta examined light captured in the Hubble deep field images, using up to 12 different light filters to separate the colors. The intensity of red was used to establish the distance to each point of light. The distances were then used to create a three-dimensional perspective of the 5,000 galaxies in the Hubble picture.

Lanzetta also used images of nearby star fields as a yardstick for stellar density and intensity to conclude that about 90 percent of the light in the very early universe was not detected by the Hubble. When this missing light was factored into the three-dimensional perspective, it showed that the peak of star formation came just 500 million years after the Big Bang and has been declining since.

Nearly 10,000 galaxies are seen in this composite image made with the Hubble Space Telescope and the deepest look, named the Hubble Ultra Deep Field, into the visible universe ever revealing a wide range of galaxies in various shapes, sizes and ages, March 9, 2004. (AP Photo/ NASA/ESA)

Current star formation, he said, "is just a trickle" of that early burst of stellar birth.

Lisa Storrie-Lombardi, a California Institute of Technology astronomer, said that the colors of the galaxies in the Hubble deep field images "are a very good indication of their distance."

Current theory suggests that about 15 billion years ago, an infinitely dense single point exploded — the Big Bang — creating space, time, matter and extreme heat. As the universe cooled, light elements, such as hydrogen and helium, formed. Later, some areas became denser with elements than others, forming gravitational centers that attracted more and more matter. Eventually, celestial bodies became dense enough to start nuclear fires, setting the heavens aglow. These were newborn stars.

Storrie-Lombardi said that current instruments and space telescopes now being planned could eventually, perhaps, see into the Dark Era, the time before there were stars.

"We are getting close to the epoch where we cannot see at all,'' she said.

Key Discovery 2: Age of Universe

Estimates of the age of the universe were revised to 13 to 14 billion years by determining the Hubble Constant and measuring the cooling rate of stars.

13 Billion Years Old

April 24, 2002

By Paul Recer

The universe is about 13 billion years old, slightly younger than previously believed, according to a study that measured the cooling of the embers in ancient dying stars.

Experts said the finding gives "very comparable results'' to an earlier study that used a different method to conclude that the universe burst into existence with the theoretical "Big Bang'' between 13 billion and 14 billion years ago.

Harvey B. Richer, an astronomer at the University of British Columbia, said the Hubble Space Telescope gathered images of the faintest dying stars it could find in M4, a star cluster some 7,000 light-years away.

Richer said the fading stars, called white dwarfs, are actually burnt out coals of stars that were once up to eight times the size of the sun. After they exhausted their fuel, the stars collapsed into Earth-sized spheres of cooling embers that eventually will turn cold and wink out of sight.

Earlier studies had established the rate of cooling for these stars, said Richer. By looking at the very faintest and oldest white dwarfs possible, astronomers can use this cooling rate to estimate the age of the universe.

Speaking at a news conference Wednesday, Richer said the dimmest of the white dwarfs are about 12.7 billion years old, plus or minus about half a billion years.

Black-and-white photo taken by the Hubble Space Telescope between January and March 2001. Peering deep inside a cluster of several hundred thousand stars, the telescope uncovered the oldest burned-out stars in our Milky Way Galaxy. Located in the globular cluster M4, these small, dying stars, called white dwarfs, are giving astronomers a fresh reading on one of the biggest questions in astronomy: How old is the universe? The ancient white dwarfs in M4 are about 12 to 13 billion years old. After accounting for the time it took the cluster to form after the big bang, astronomers found that the age of the white dwarfs agrees with previous estimates for the universe's age. Released April 24, 2002. (AP Photo/NASA and H. Richer, University of British Columbia)

Richer said it is estimated that star formation did not begin until about a billion years after the Big Bang. He said this means his best estimate for age of the universe is "about 13 billion years."

Three years ago, astronomers using another method estimated the age at 13 to 14 billion years. That was based on precise measurements of the rate at which galaxies are moving apart, an expansion that started with the Big Bang. They then back-calculated — like running a movie backward — to arrive at the age estimate.

www.ingramcontent.com/pod-product-compliance
Lightning Source LLC
Chambersburg PA
CBHW050404120526
44590CB00015B/1820